MW00451215

SATURN IB / SATURN V

ROCKET PAYLOAD PLANNER'S GUIDE

MISSILE & SPACE SYSTEMS DIVISION
DOUGLAS AIRCRAFT COMPANY, INC.
SANTA MONICA/CALIFORNIA

ISBN #978-1-937684-77-8
www.PeriscopeFilm.com

DOUGLAS

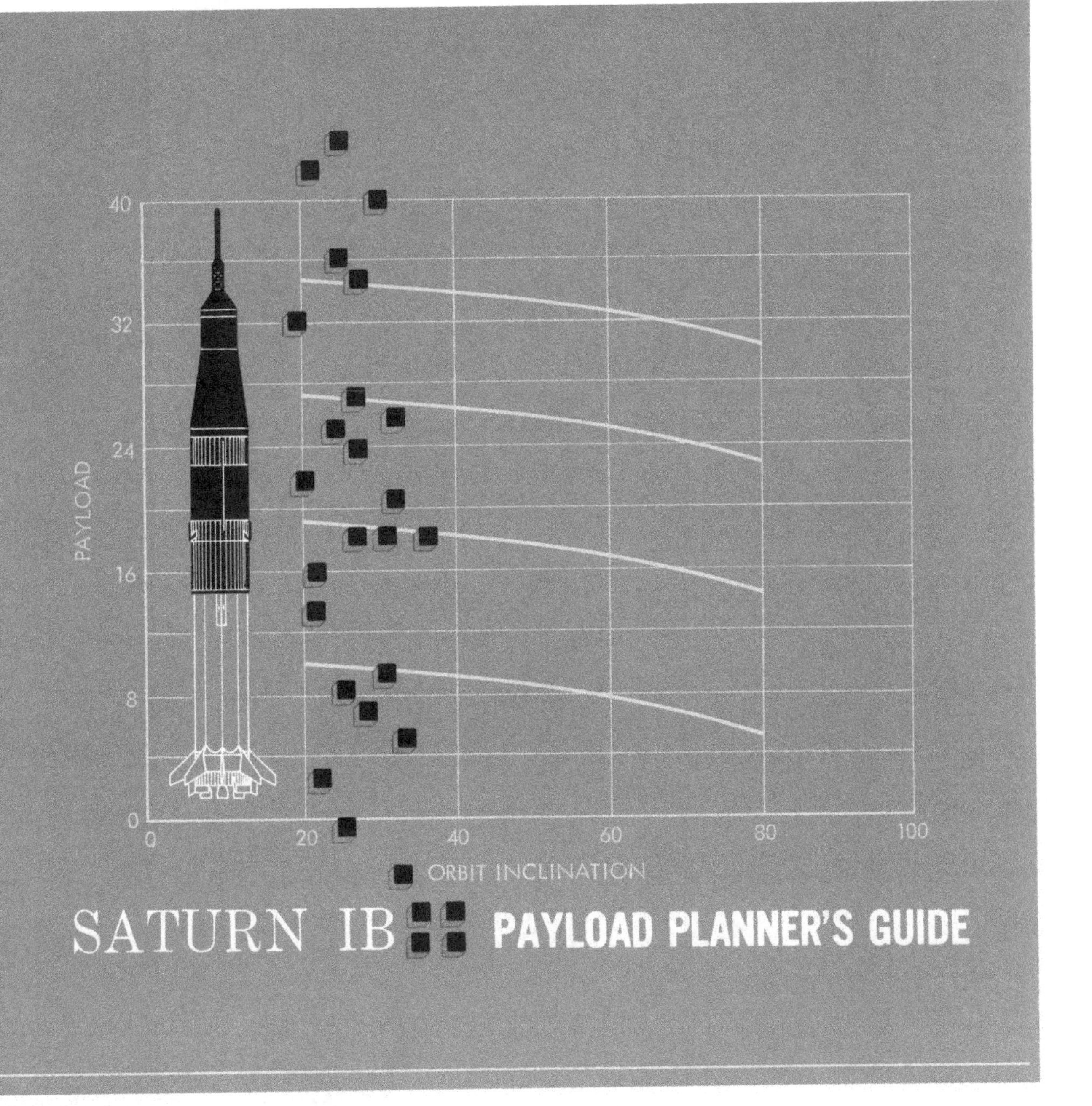

MISSILE & SPACE SYSTEMS DIVISION
DOUGLAS AIRCRAFT COMPANY, INC.
SANTA MONICA/CALIFORNIA

SATURN IB

PAYLOAD PLANNER'S GUIDE

June 1965

Douglas Report SM-47010

Prepared By: L. O. Schulte
D. D. Wheeler

Program Planning - Saturn
Payload Applications

Approved By: F. C. Runge
Program Manager - Saturn
Payload Applications

Approved By: T. J. Gordon
Director of Advance Saturn and
Large Launch Vehicles

DOUGLAS MISSILE & SPACE SYSTEMS DIVISION
HUNTINGTON BEACH, CALIFORNIA

FOREWORD

This book has been prepared by Douglas to acquaint payload planners with the capability of the Saturn IB Launch Vehicle and to assist them in their initial payload/launch vehicle planning. This guide is not an offer of space aboard Saturn. Only NASA can commit experiments to this vehicle. This book attempts to show methods by which Saturn could accommodate payloads of various weights, volumes and missions. You will see that the capabilities of this vehicle permits a wide spectrum of assignments, including scientific, engineering, technological as well as numerous operational type payloads.

Requests for additional information may be addressed to:

Mr. Fritz Runge, Program Manager
Saturn Payload Applications
Douglas Missile and Space Systems Division
5301 Bolsa Avenue
Huntington Beach, California 92646

Telephone 714-897-0311

SATURN IB PAYLOAD PLANNER'S GUIDE

TABLE OF CONTENTS

SATURN IB ON PAD

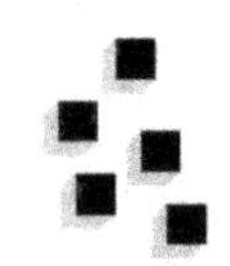

The Saturn IB is a two stage launch vehicle being developed by NASA to achieve many objectives related to the national goal of lunar exploration and other space flights. The Saturn IB vehicle is designed to perform large payload, manned and unmanned space flights economically and with reliability already demonstrated by the significant success of Saturn I flights. The Saturn IB vehicles are scheduled for initial flight tests during 1965 and 1966 and will feature an operational payload capability of over 35,000 lb in a 100 nautical mile circular earth orbit. Early Saturn IB flights are oriented toward the development of the Apollo system capability. However, subsequent to these flights the Saturn IB can be used for other large prime payloads or smaller auxiliary payloads. Advanced two-stage Saturn IB launch vehicles and three-stage Saturn IB/Centaurs are in planning. These vehicles will extend the Saturn IB mission potential to reach even higher energy orbits, including lunar and interplanetary transfer trajectories.

This Payload Planner's Guide is intended as a starting point for engineers, scientists, and executives who want to plan and conduct engineering tests or experiments in the space sciences or consider operational mission possibilities. It outlines for the payload planner the technical description and the procedures with which large prime, or small auxiliary payloads can be effectively integrated and flown on the vehicle. The payload planner will find here the characteristics of the Saturn IB launch vehicle, its performance, the accommodations it offers to potential experimenters, suggested procedures to be followed in obtaining support for the experiment, approximate flight schedules and engineering data needed to initiate the design of a payload. To planners of prime payloads, the guide offers three protective shroud designs. To auxiliary payload planners, it presents the proposed Douglas SPACE-PAC concept identifying and describing several places in the Saturn IB where such payloads could be carried. Environmental data and payload weight limitations for each payload volume are provided.

The Saturn IB performance capabilities are included for payload flight planning. The major subsystems of the launch vehicle and their relation to the payloads are described. A concept-to-flight chronology of events is presented to support payload/launch vehicle system planning on the part of prospective users.

The Douglas Missile and Space Systems Division will be pleased to discuss the planning, support, operation, and data evaluation involved in the flight of any payload on Saturn IB.

SATURN IB CONFIGURATIONS

The two and three stage configurations of the Saturn IB, depicted in Figure I-1 are the basis for the data presented in this guide. The two stage Saturn IB with the Apollo Spacecraft is 223 feet tall and weighs nearly 625 tons. The three-stage Saturn IB/Centaur with a Voyager payload is 199 feet high and weighs over 640 tons.

FIRST STAGE (S-IB)

The S-IB stage incorporates tanks developed and qualified on the Redstone, Jupiter, and Saturn I vehicles. Nine such tanks are used in the S-IB stage; five for LOX and four for RP-1 (Kerosene). Eight 200,000 lb thrust Rocketdyne H-1 engines are used. By clustering eight of these engines, a thrust of 1.6 million pounds is generated at lift-off, to propel the entire vehicle to an altitude of approximately 30 n.mi. The outer ring of engines are hydraulically gimballed to provide thrust vector control in response to steering commands from the guidance system located in the Instrument Unit. The first stage is separated from the second, by four Thiokol solid rocket motors, each of which produces 36,000 lb of thrust for 1.5 seconds.

SECOND STAGE (S-IVB)

The S-IVB stage is powered by a single Rocketdyne J-2 engine that burns liquid oxygen and liquid hydrogen to provide a thrust of 200,000 lb. During flight, control signals from the Instrument Unit provide pitch and yaw control by gimballing the engine. Roll control is provided by 150 lb thrust engines located in the Auxiliary Propulsion System (APS) modules. Three axis (roll, pitch and yaw) attitude control during coast is provided entirely by the APS. After first stage separation, the J-2 engine on the S-IVB stage ignites and propels the payload to the desired altitude. Three solid-propellant ullage rockets are fired at stage separation, just prior to ignition of the J-2 engine, to assure that the propellant is settled in the bottom of the tanks during the start phase.

Figure I-1
SATURN IB TWO AND THREE STAGE CONFIGURATIONS

INSTRUMENT UNIT

The Instrument Unit houses the guidance and control systems and the flight instrumentation systems for the Saturn IB launch vehicle. This Unit constitutes a structural link between the S-IVB stage and the payload or third stage. It carries electrical, guidance and control, instrumentation, measuring, telemetry, radio frequency, environmental control, and emergency detection systems.

THIRD STAGE (CENTAUR)

A Saturn IB three-stage configuration uses the flight-proven Centaur as a third stage. The Centaur uses two 15,000 lb thrust restartable Pratt and Whitney RL-10A-3 LOX/LH_2 engines. It is separated from the Instrument Unit, S-IVB stage by four 3,400 lb thrust retro-rockets. Centaur has its own guidance and control system which controls the stage after its separation from the S-IVB.

A more detailed description of the two and three stage Saturn IB configurations and the individual stages is presented in Section III.

SATURN IB CAPABILITY

Saturn IB has the capability to perform a broad spectrum of manned and unmanned space missions carrying large prime and auxiliary payloads as summarized in Figure I-2 and presented in detail in Section IV.

The major advantages of the Saturn IB that may be of interest to auxiliary and prime payload planners are:

- High orbital payload capability (about 35,000 lb to 100 n.mi.).
- High altitude capability (about 10,000 lb to 20,000 n.mi.).
- Low transportation costs per pound of payload in orbit.
- Flight proven systems and subsystems.
- Man-rated systems.
- Production stages.
- Complete and existing manufacturing, test and launch facilities.
- Flexibility of planning two, three or four stage configurations.
- Associated NASA data acquisition and tracking networks are operational.
- Auxiliary payload volumes, weight, power, and data channels may be available.
- Growth potential of vehicle is considerable.

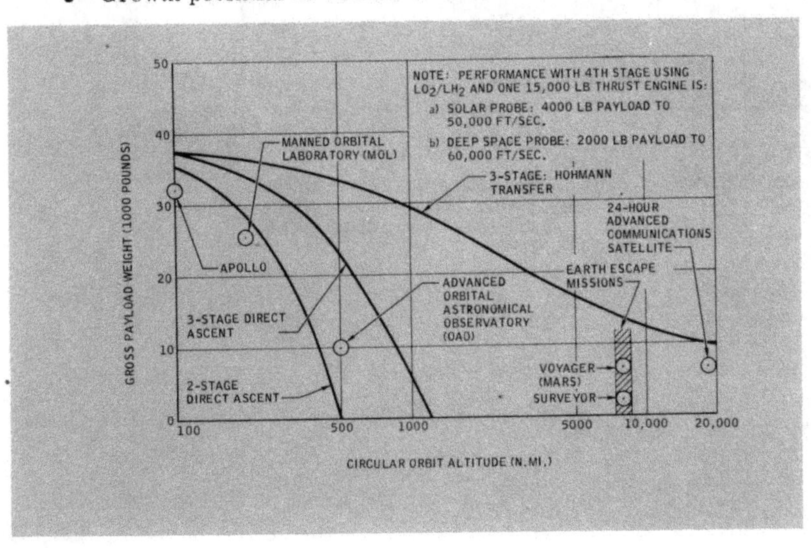

Figure I-2
SATURN IB MISSION POTENTIAL

Figure II-1
SATURN IB PAYLOAD POTENTIAL

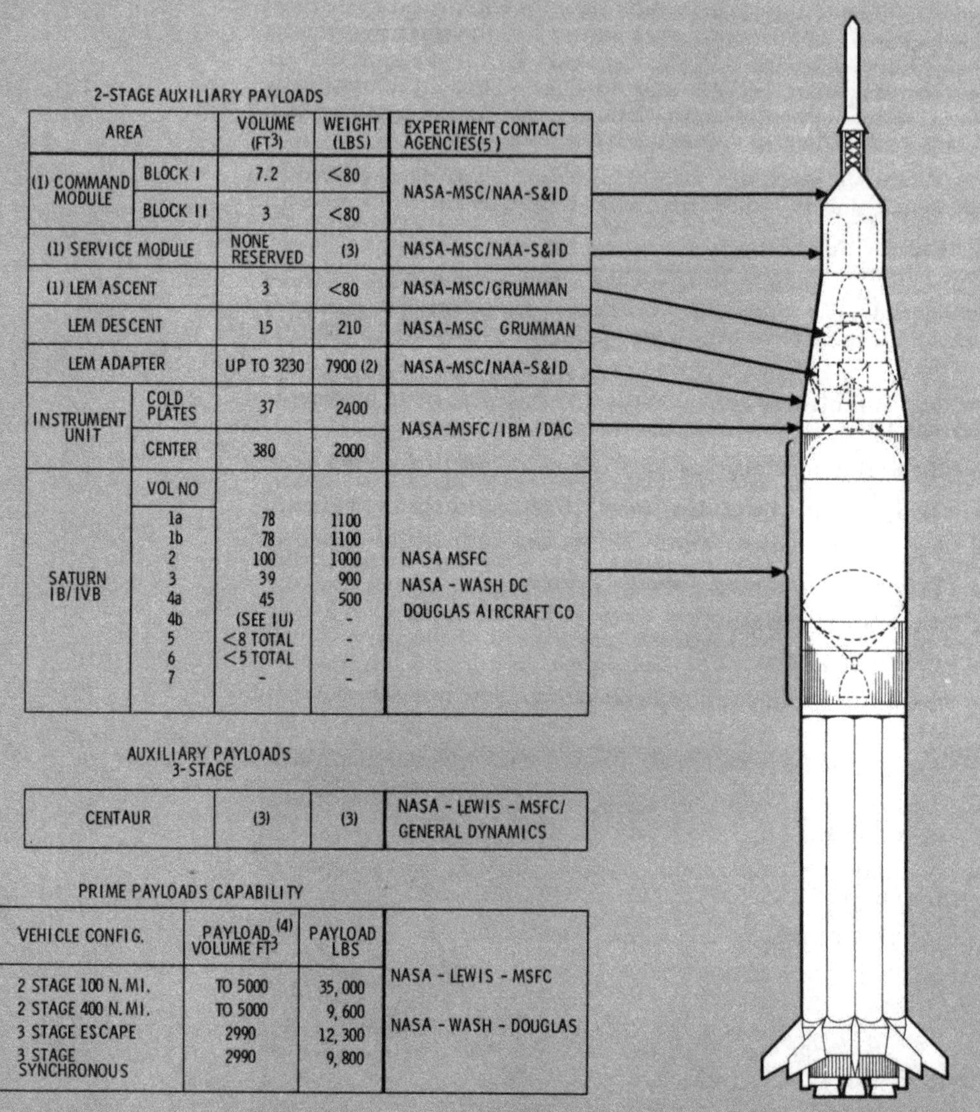

2-STAGE AUXILIARY PAYLOADS

AREA		VOLUME (FT^3)	WEIGHT (LBS)	EXPERIMENT CONTACT AGENCIES(5)
(1) COMMAND MODULE	BLOCK I	7.2	<80	NASA-MSC/NAA-S&ID
	BLOCK II	3	<80	
(1) SERVICE MODULE		NONE RESERVED	(3)	NASA-MSC/NAA-S&ID
(1) LEM ASCENT		3	<80	NASA-MSC/GRUMMAN
LEM DESCENT		15	210	NASA-MSC GRUMMAN
LEM ADAPTER		UP TO 3230	7900 (2)	NASA-MSC/NAA-S&ID
INSTRUMENT UNIT	COLD PLATES	37	2400	NASA-MSFC/IBM/DAC
	CENTER	380	2000	
SATURN IB/IVB	VOL NO			NASA MSFC NASA - WASH DC DOUGLAS AIRCRAFT CO
	1a	78	1100	
	1b	78	1100	
	2	100	1000	
	3	39	900	
	4a	45	500	
	4b	(SEE IU)	-	
	5	<8 TOTAL	-	
	6	<5 TOTAL	-	
	7	-	-	

AUXILIARY PAYLOADS
3-STAGE

CENTAUR	(3)	(3)	NASA - LEWIS - MSFC/ GENERAL DYNAMICS

PRIME PAYLOADS CAPABILITY

VEHICLE CONFIG.	PAYLOAD(4) VOLUME FT^3	PAYLOAD LBS	
2 STAGE 100 N. MI.	TO 5000	35,000	NASA - LEWIS - MSFC
2 STAGE 400 N. MI.	TO 5000	9,600	
3 STAGE ESCAPE	2990	12,300	NASA - WASH - DOUGLAS
3 STAGE SYNCHRONOUS	2990	9,800	

(1) NPC 500-9 APOLLO IN-FLIGHT EXPERIMENT GUIDE DATED SEPT 15, 1964.

(2) EQUAL TO TOTAL LEM WEIGHT.

(3) SEE CONTACT AGENCIES

(4) FINAL AUXILIARY PAYLOAD WEIGHT AND VOLUME DEPENDS ON PRIME MISSION

(5) INFORMATION ON EXPERIMENT SUBMITTAL PROCESS AND ASSOCIATED VEHICLE DATA CAN BE OBTAINED FROM COGNIZANT NASA AGENCIES.

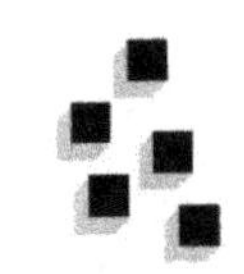

A. Planning and Schedules

Technical assistance is available at Douglas to aid the experimenter or payload originator in planning, flying and evaluating a payload on the Saturn IB. Two-stage and three-stage Saturn IB vehicles can carry prime or auxiliary payloads on a great variety of manned or unmanned missions. It is beyond the scope of this guide to include all the data on each payload volume. Some of the significant examples are shown in Figure II-1. Space, power and weight carrying capability is available in almost every part of the vehicle. Depending on specific mission requirements auxiliary payloads may be carried in the:

1) Apollo Command Module
2) Apollo Service Module
3) Lunar Excursion Module (LEM) Ascent or Descent Modules
4) LEM Adapter
5) Instrument Unit (I.U.)
6) S-IVB Stage
7) Centaur Stage (in three-stage configurations)

New prime payloads may be selected for either two- or three-stage missions using existing or special shroud designs. A summary of Saturn IB payload potential is presented in Figure II-1.

Information in this handbook is primarily associated with the prime payload carrying ability of the Saturn vehicle, auxiliary payloads within the S-IVB and payloads supported by the S-IVB and extending into the I. U. Experimenters desiring more information on the other stages or modules can find the appropriate agency to contact in Figure II-1. Discussions in this guide are confined to the prime payloads on Saturn IB and auxiliary payloads in the S-IVB and the I. U.

The steps normally required to bring a prime or auxiliary payload from concept, through integration and flight with the launch vehicle system, to final evaluation, are presented in the flow-diagram shown in Figure II-2.

Payloads and experiments may be conceived by many organizations and individuals in government, universities and industry. In some cases, to be effective, payload proposals must include certain vehicle and program interface data. Douglas will assist these experiment originators in definition of a payload/launch vehicle concept. The proposals submitted by the originating organization, will then be evaluated by NASA experiment review boards to determine the concept's priority in meeting national objectives. With mission objectives approved, budget and vehicle allocations can be made and the concept can be processed through normal procurement channels to obtain the final contractual authority. The Saturn Payload Applications organization of the Douglas Missile and Space Systems Division can supply planning information in support of payload proposals.

Upon receipt of contractual authority, more detailed mission planning will be accomplished among NASA, the Saturn IB stage contractors, and the payload originator. NASA acts as overall program integration manager.

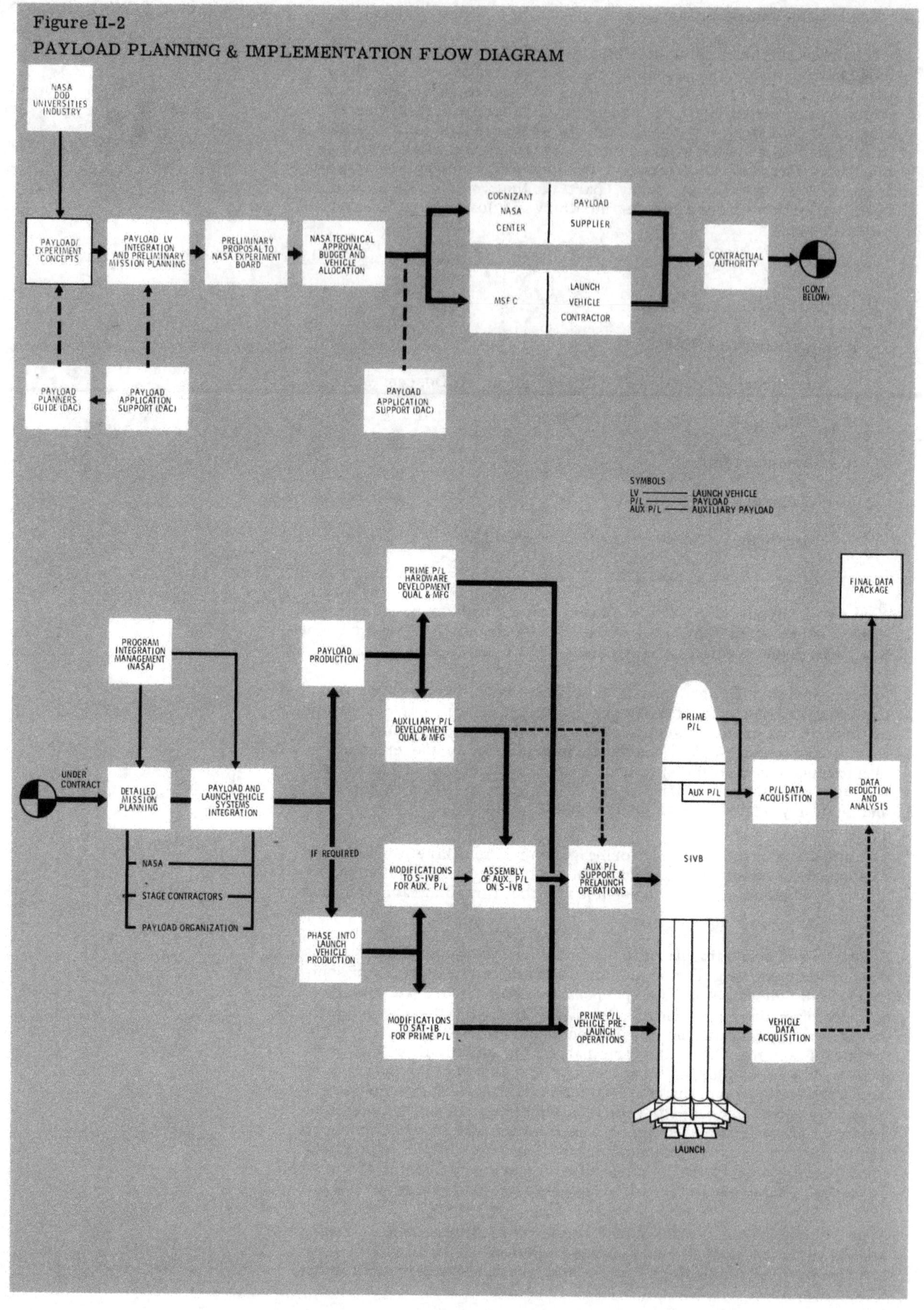
Figure II-2
PAYLOAD PLANNING & IMPLEMENTATION FLOW DIAGRAM
NASA DOD UNIVERSITIES INDUSTRY
PAYLOAD/ EXPERIMENT CONCEPTS
PAYLOAD LV INTEGRATION AND PRELIMINARY MISSION PLANNING
PRELIMINARY PROPOSAL TO NASA EXPERIMENT BOARD
NASA TECHNICAL APPROVAL BUDGET AND VEHICLE ALLOCATION
COGNIZANT NASA CENTER
PAYLOAD SUPPLIER
MSFC
LAUNCH VEHICLE CONTRACTOR
CONTRACTUAL AUTHORITY
(CONT BELOW)
PAYLOAD PLANNERS GUIDE (DAC)
PAYLOAD APPLICATION SUPPORT (DAC)
PAYLOAD APPLICATION SUPPORT (DAC)
SYMBOLS
LV — LAUNCH VEHICLE
P/L — PAYLOAD
AUX P/L — AUXILIARY PAYLOAD
PRIME P/L HARDWARE DEVELOPMENT QUAL & MFG
FINAL DATA PACKAGE
PROGRAM INTEGRATION MANAGEMENT (NASA)
PAYLOAD PRODUCTION
AUXILIARY P/L DEVELOPMENT QUAL & MFG
PRIME P/L
UNDER CONTRACT
DETAILED MISSION PLANNING
PAYLOAD AND LAUNCH VEHICLE SYSTEMS INTEGRATION
AUX P/L
P/L DATA ACQUISITION
DATA REDUCTION AND ANALYSIS
IF REQUIRED
SIVB
NASA
STAGE CONTRACTORS
PAYLOAD ORGANIZATION
MODIFICATIONS TO S-IVB FOR AUX. P/L
ASSEMBLY OF AUX. P/L ON S-IVB
AUX P/L SUPPORT & PRELAUNCH OPERATIONS
PHASE INTO LAUNCH VEHICLE PRODUCTION
MODIFICATIONS TO SAT-IB FOR PRIME P/L
PRIME P/L VEHICLE PRE-LAUNCH OPERATIONS
VEHICLE DATA ACQUISITION
LAUNCH

Development and qualification of payloads proceeds in parallel with launch vehicle production. Peculiar payload requirements may necessitate accomplishment of detailed testing, test support planning, and test documentation must be accomplished at the beginning of final checkout of the Saturn vehicle to insure compatibility of payloads and launch vehicle.

A typical schedule of S-IVB stage production and of critical auxiliary payload integration periods is shown in Figure II-3. Also shown are typical delivery dates that an auxiliary payload would have to meet to minimize interference with the delivery schedule of the main stages. The complexity of the payload and the nature of its integration will establish the lead time for a particular flight.

A schedule indicating the type of operations that must be accomplished at Kennedy Space Center (KSC) to prepare a prime payload is presented in Figure II-4.

Figure II-5 indicates typical delivery dates to KSC for Saturn IV vehicles maybe estimated at a rate of six per year. Many of these have prime payload assignments. However, some are not expected to be fully loaded and may have room for auxiliary payloads.

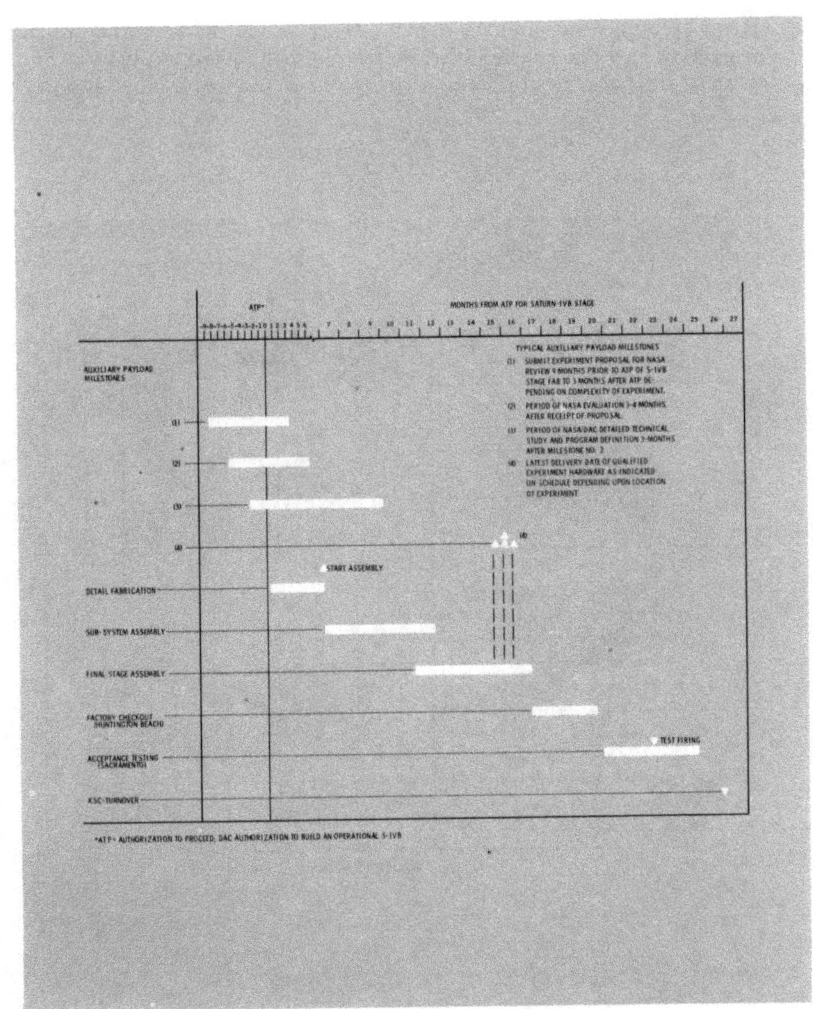

Figure II-3
TYPICAL S-IVB INTEGRATION SCHEDULE FOR AUXILIARY PAYLOADS

B. Launch Vehicle Accommodations

Since auxiliary payloads can vary widely in size, shape and weight, the S-IVB stage has been reviewed in considerable detail to find possible locations in which auxiliary payloads can be carried if a weight allowance is available on a flight. At least seven volumes may be used depending upon the experimenter's specific requirements.

1. Auxiliary Payloads Mounted in S-IVB Stage

Convenient volumes may be available to experimenters in the forward portion of the stage and in the I.U. The envelopes of available space within the forward skirt and I.U. extends from the electrical/electronic units mounted on the skirt to the forward dome of the S-IVB tank as shown in Figure II-6. Also, additional space is available in pods mounted externally on the forward skirt. Many combinations of space, power, data, and environmental services can be furnished to meet the needs of auxiliary payloads. These systems do not exist in the present vehicle. This discussion is intended to illustrate feasible techniques which could be employed to accommodate auxiliary payloads.

The proposed experiment volumes of the Saturn S-IVB stage included in the Douglas "SPACE-PAC concept" are as follows:

Experiment Volumes No. 1a and 1b (Figure II-6)

About 78 cubic feet could be provided external to the forward interstage in each of two pods as shown. Since these pods have not been designed, it may be possible to include provisions for certain unique payload

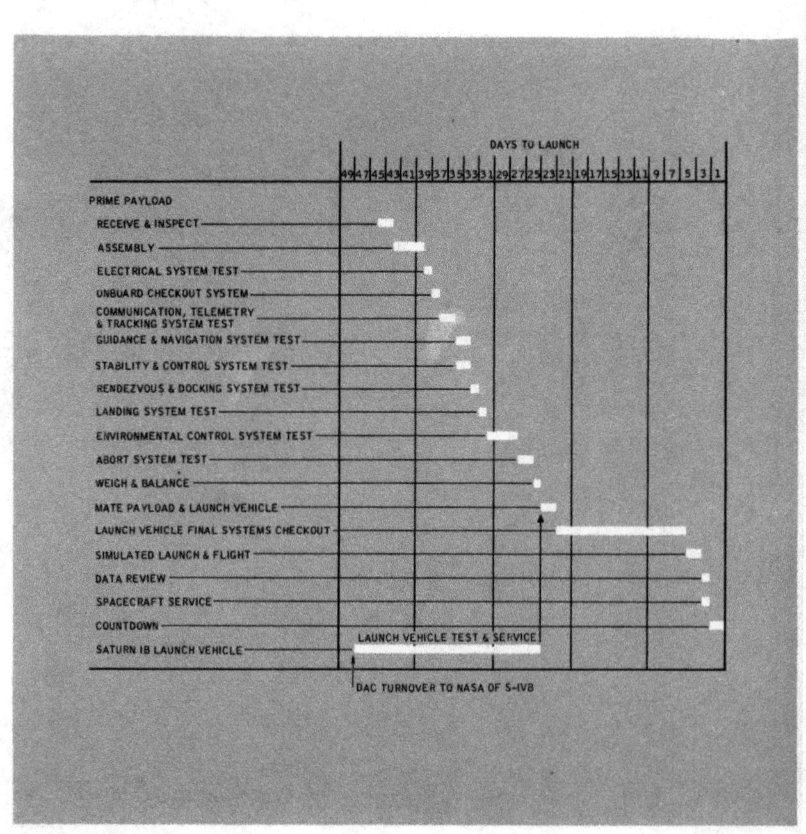

Figure II-4

TYPICAL KSC SATURN PREPARATION SCHEDULE FOR PRIME PAYLOAD

Figure II-5

SATURN IB LAUNCH VEHICLE DELIVERY SCHEDULE TO KSC

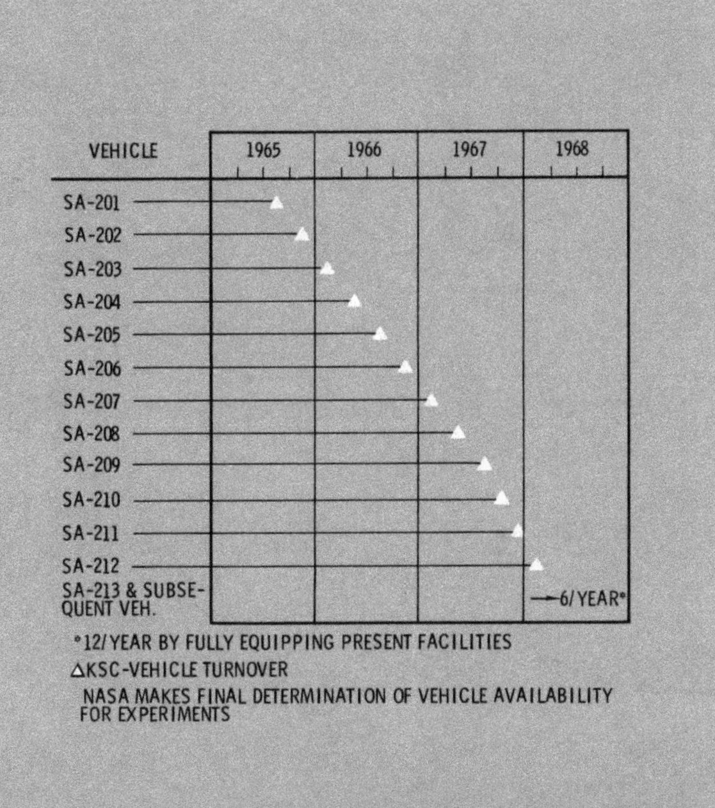

requirements in the basic layout of these volumes. Approximately 1,100 lb of payload may be carried in each location. Some modification to the forward skirt for structural support and rerouting of some electrical cables will be required. Mounting concepts for these experiment volumes are shown in Figure II-7. The first concept shows a payload incorporating a solid propellant motor (ABL-X-258), an optional second motor (ARC-XM-85), and a satellite payload. The assembly is mounted in a support cradle that also provides means of ejection from the S-IVB stage. The ABL-X-258 motor is ignited by a signal from a timer keyed to the ejection sequence. The payload assembly has an attitude control system referenced to the S-IVB stage attitude at payload separation. The payload is protected through the boost phase of the trajectory by a fairing that is jettisoned just before payload ejection. The support cradle is attached to a honeycomb mounting plate that in turn is attached to the S-IVB stage structure through the forward skirt frames and to pads on the S-IVB Liquid Hydrogen (LH_2) tank skin. Some performance figures for this type of installation are shown in Configuration B Figure II-8. Other payloads with different requirements that could also be accommodated are indicated by concepts 2, 3, and 4 of Figure II-7.

Experiment Volume No. 2

Variation in the shape of Volume 2 is possible depending on the payload configuration. However, some limitations on the use of

Figure II-6

PROPOSED SPACE-PAC CONCEPT S-IVB AUXILIARY PAYLOAD VOLUMES

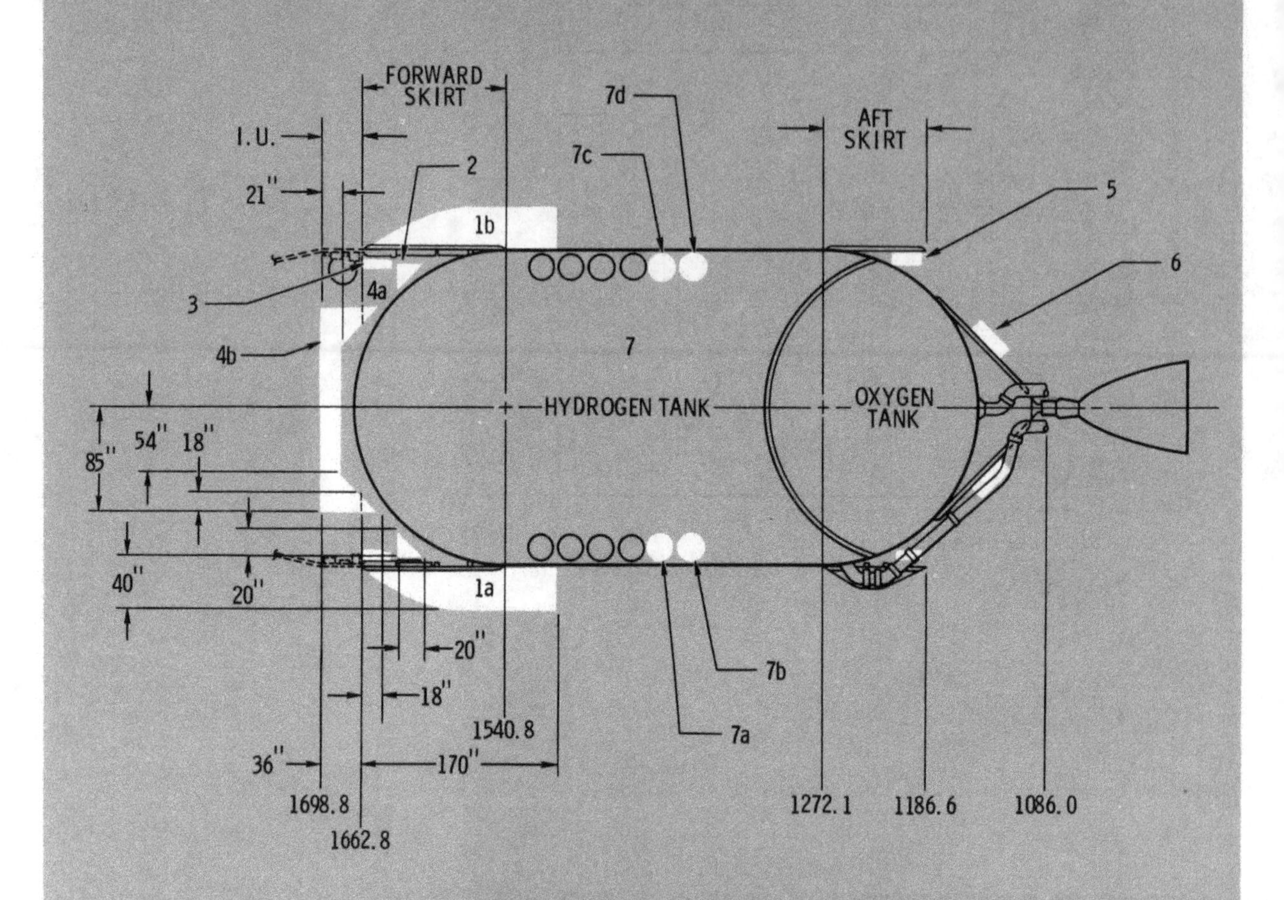

VOL. NO.	LOCATION	VOLUME	PAYLOAD WEIGHT, LBS
1a	FWD. SKIRT - EXT.	78 FT^3	1,100
1b	FWD. SKIRT - EXT.	78 FT^3	1,100
2	FWD. SKIRT - INT.	109 FT^3	1,000
3	FWD. SKIRT - INT.	39 FT^3	900
4a	FWD. SKIRT - INT.	45 FT^3	2,500
4b	I.U. - INT.	380 FT^3	
5	AFT SKIRT - INT.	8 FT^3	-
6	THRUST STRUCTURE	INDEF.	-
7	HYDROGEN TANK	-	-
7a-d	HYDROGEN TANK	3.5 FT^3 EA.	-

Figure II-7

S-IVB FORWARD SKIRT POD CONFIGURATION (VOLUMES I-A & I-B)

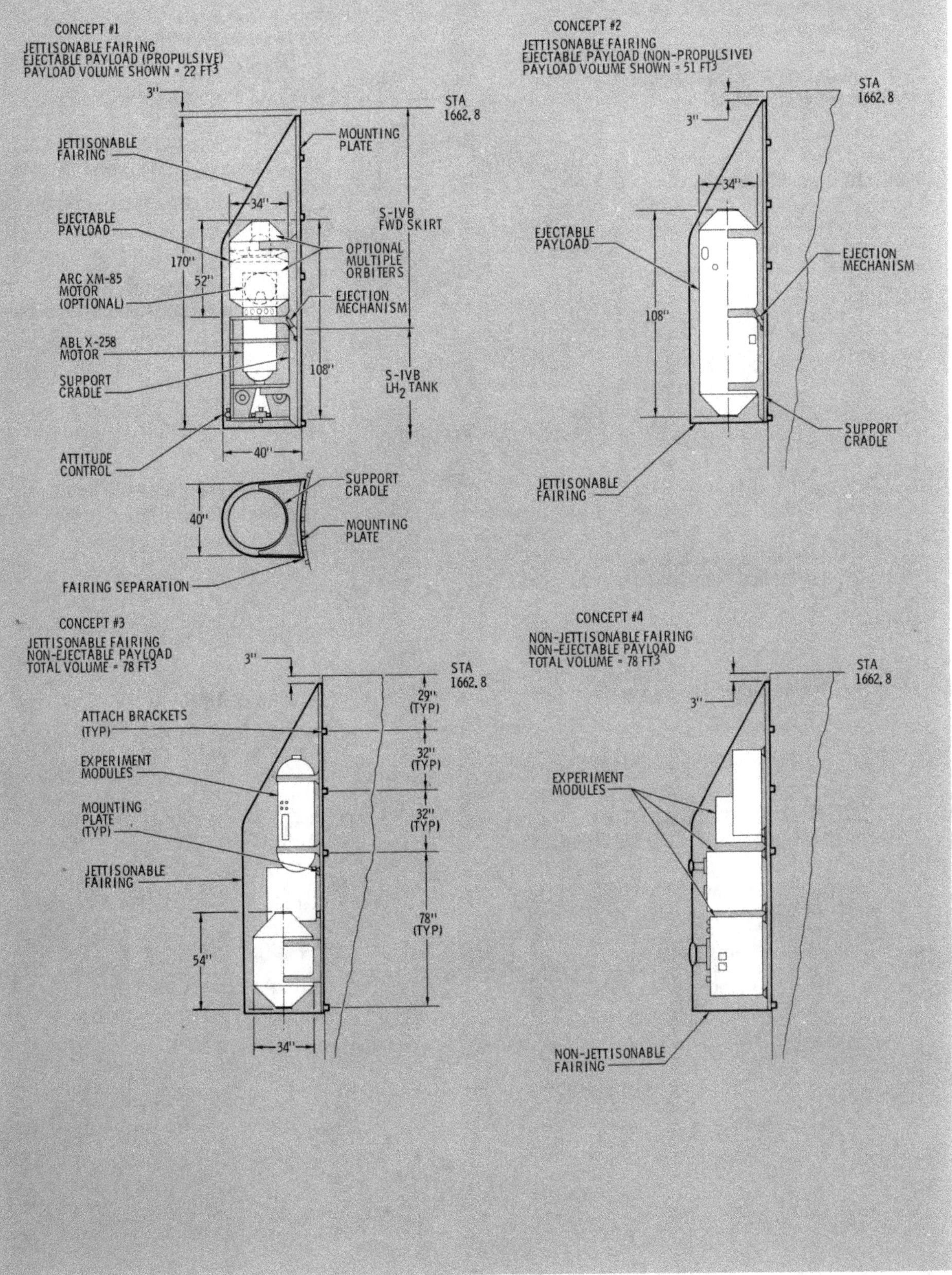

Figure II-8

SATURN S-IVB STAGE SPACE-PAC ALTERNATE CONFIGURATIONS

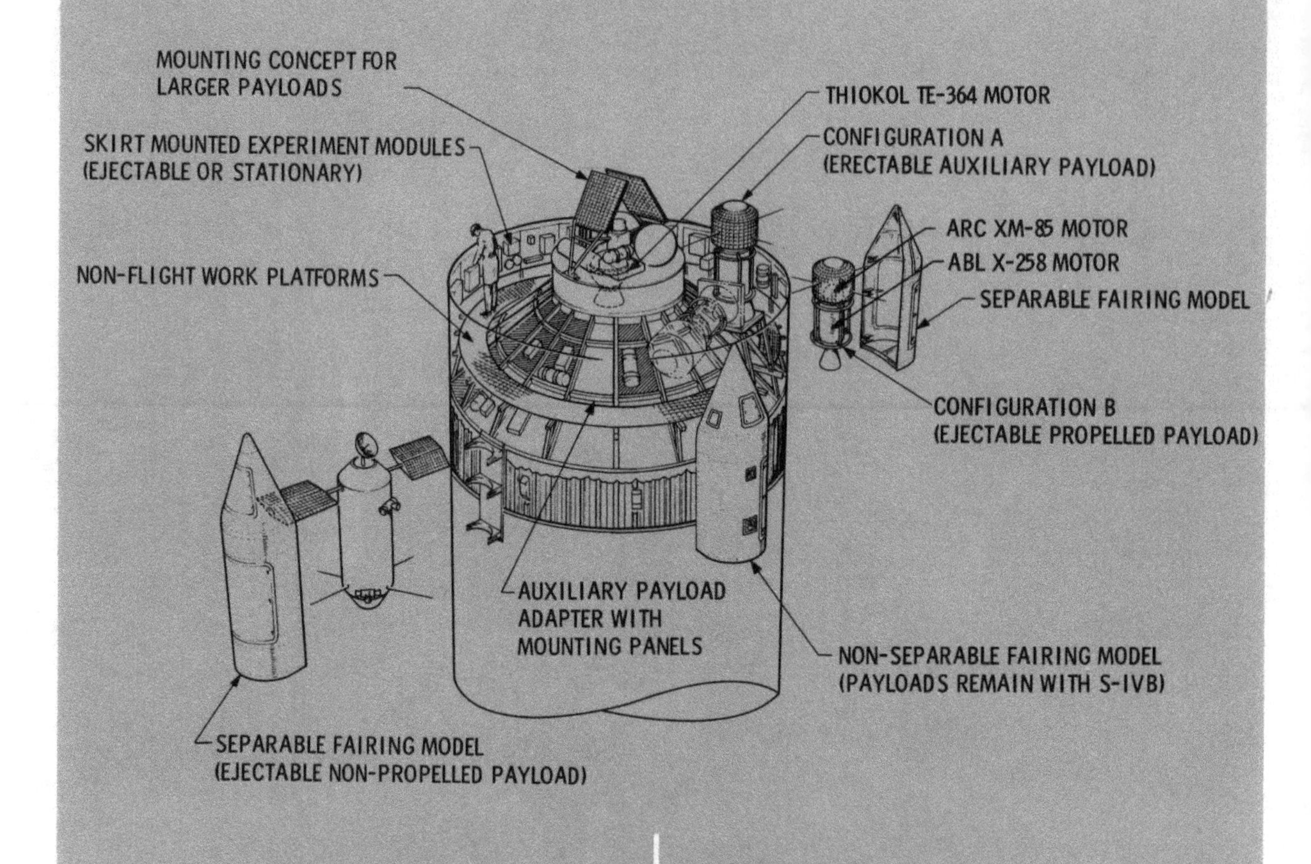

CONFIGURATION A
(ABL X-258 MOTOR + SATELLITE)

ELLIPTICAL ORBIT CAPABILITIES
(INITIAL ORBIT ALTITUDE = 100 N. MILES)

SATELLITE WEIGHT (LBS)*	APOGEE (N. MI.)	PERIGEE (N. MI.)
157	LUNAR MISSION	—
200	44,000	100
300	15,300	100
400	9,500	100
500	7,000	100

CONFIGURATION B
(ABL X-258 MOTOR + ARC XM-85 MOTOR + SATELLITE)

CIRCULAR ORBIT CAPABILITIES**
(INITIAL ORBIT ALTITUDE = 100 N. MI.)

SATELLITE WEIGHT (LBS)*	ORBITAL ALTITUDE (N. MI.)
150	7,700
200	4,700
300	2,850

ELLIPTICAL ORBIT CAPABILITIES
(INITIAL ORBIT ALTITUDE = 100 N.M.)

SATELLITE WEIGHT (LBS)*	APOGEE (N. MI.)	PERIGEE (N. MI.)
100	24,700	17,000
200	9,800	6,600
300	7,000	3,500

*SATELLITE WEIGHT INCLUDES GUIDANCE SYSTEM

**REQUIRES OFF-LOADING X-258 MOTOR

this space are set by checkout requirements on equipment and wiring in the interstage, I.U., and the LEM descent module. Accessibility to these areas requires the use of a vertical access kit that restricts the available volume to that under the access kit platform. This volume consists of approximately 109 cubic feet. The experiment modules can be mounted on a lightweight structural cone supported by one of the forward skirt frames. A total payload weight of about 1,000 lb can be carried in this location. Weight limitations on a specific experiment module must be controlled by prime mission requirements as well as by structural design factors.

Experiment Volume No. 3

Experiment modules can be mounted directly on the thermal conditioning panels in the forward skirt. See Figure II-9. On all vehicles after SA-205, six of the sixteen panels will be available for mounting experiments because of a simplified telemetry system. A volume of at least 39 cubic feet with a maximum weight of 900 lb is available. This weight and volume indicated may be increased if the accessibility, the payload center of gravity location and the mounting method permit.

Experiment Volume No. 4

For some missions in which the LEM descent stage is not carried, an additional large volume may be available above the access kit platform. This space extends over the S-IVB forward dome and into the instrument unit to Station 1698.8. The volume available within the forward skirt is about 45 cubic feet. Approximately 380 cubic feet is available in the I.U. The experiment modules can be mounted on an auxiliary payload adapter. This payload adapter consists of "spider" structure supported from the S-IVB forward skirt frames as shown in Figure II-10. The adapter would also serve as an access kit

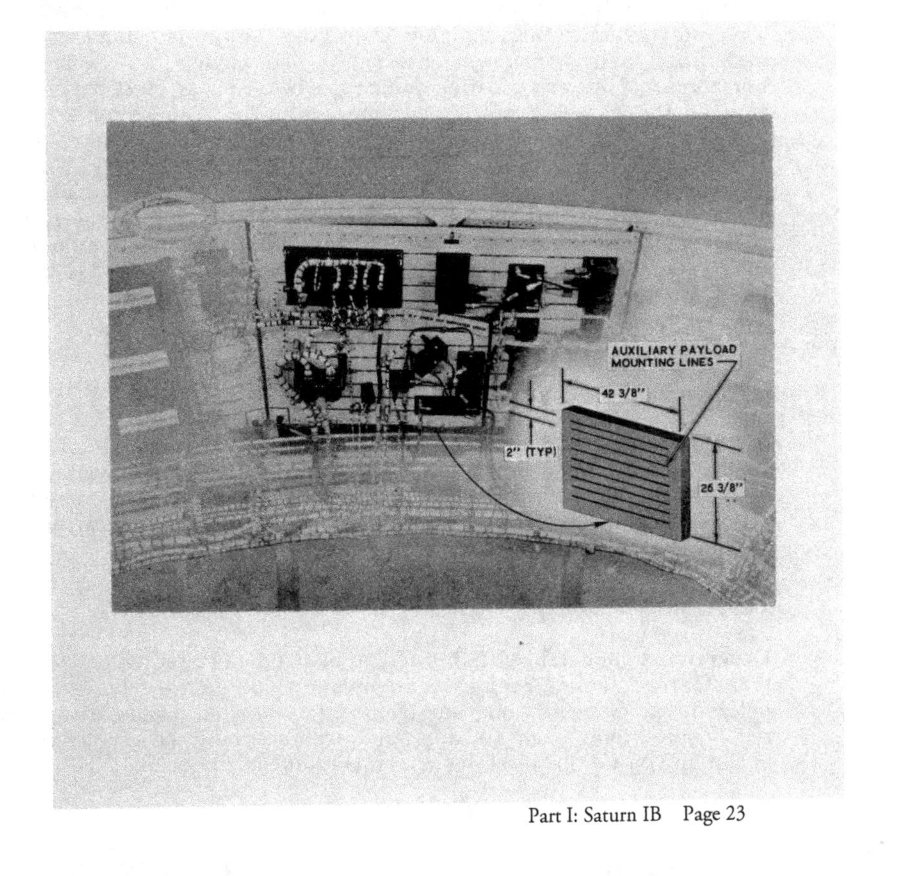

Figure II-9
S-IVB FORWARD
SKIRT THERMAL
CONDITIONED PANELS

Figure II-10
AUXILIARY PAYLOAD ADAPTER

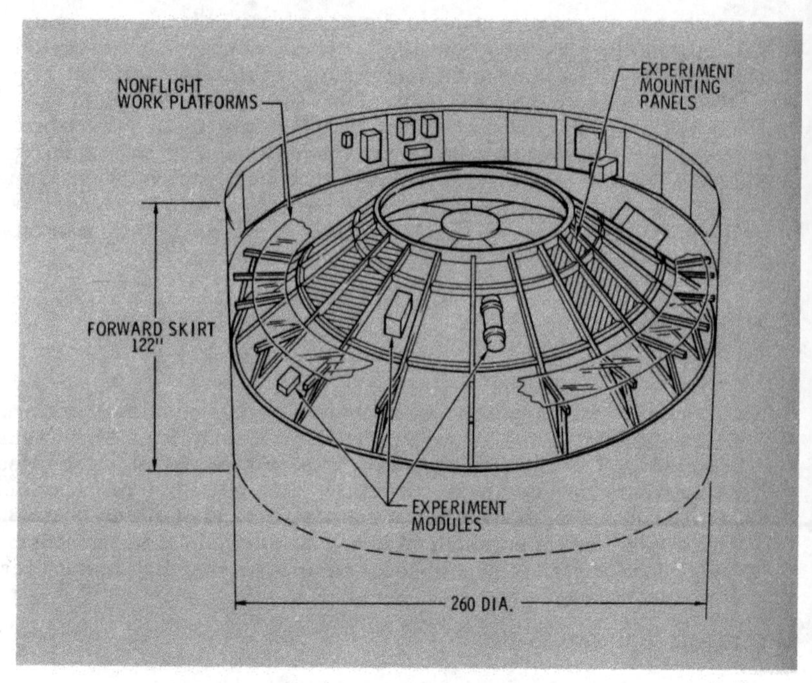

when removable work platforms are inserted as shown. The experiment modules are mounted on honeycomb panels attached to the adapter. The adapter and accompanying experiment modules must be removable for access to the liquid hydrogen tank. A total payload weight of up to 2,500 lb may be carried in this location, prime payload weight permitting.

Other auxiliary payloads, such as the Delta third stage, may be carried as shown in Figure II-8. The payloads are mounted on the auxiliary payload adapter through additional supporting structure. The internal mounting depicted shows the Delta third stage including separation and spin-up mechanism. The payload is carried in a horizontal or stowed position during the boost phase until the separation of the S-IVB from the prime payload. At this time the Delta third stage is erected, spun-up, and separated at a signal in the S-IVB stage separation sequence. Ignition of the ABL-X-258 motor is triggered by a timer after an appropriate separation distance is achieved. Separation forces can be generated by small solid propellant motors similar to the spin rockets. In the stowed position the Delta third stage projects approximately 33 inches above the S-IVB/I.U. interface at Station 1662.8. Some representative performance figures for two possible configurations are shown.

Experiment Volume No. 5

A small amount of usable volume may be available in the aft skirt area of the operational configuration. Certain experiment modules (five volumes of about 1.5 cubic feet each) may be mounted directly on the existing mounting plates in place of R&D equipment not required on operational flights.

Experiment Volume No. 6

Experiment modules of light weight may be mounted directly on the thrust structure. Precise locations and volumes available cannot be defined at present, but small modules of the proper size (about one cubic foot each), shape, and weight could be accommodated depending on the mounting requirements of the payloads.

Experiment Volume No. 7

Volume 7 is within the hydrogen tank itself. Any experiment placed in this volume would, of course, displace the LH_2 and be subjected to the temperature and pressure conditions of the LH_2. Some experimenters may want to take advantage of these conditions to study a system under cryogenic and space environment, or to study the fluid behavior of the liquid or gaseous hydrogen. There are eight cold helium spheres in the LH_2 tank to pressurize the liquid oxygen (LOX) tank during powered flight. There are four additional flanged connections on which spheres could be installed to hold experiments at liquid hydrogen temperature while protecting them from direct contact with the hydrogen. Each of the spheres has a volume of 3.5 cubic feet. The entrance to the sphere is only 1.44 inches in diameter. However, this could be increased to about 4 inches in diameter.

2. Prime Payloads Above the S-IVB Stage

Minimum vehicle changes will be required if the volume normally occupied by the Lunar Excursion Module (LEM) were to be utilized for other payloads (Figure II-11). This volume within the LEM adapter might be used on future flights if prime mission objectives permit.

Figures II-12 and II-13 illustrate two other possible configurations of payloads and payload fairings. The fairings are shown as on a two stage Saturn IB but they are equally suitable for the three stage configuration. The fairings protect the payloads from aerodynamic loads and temperatures while in flight through the atmosphere. They may also be used for payload thermal conditioning on the launch pad, if such conditioning is required and provisions for it are included. The fairings may be made of aluminum honeycomb or of fiberglass if RF transparency is a requirement. The fairings are jettisoned during second stage operation when atmospheric effects are negligible.

A typical adapter which supports the prime payload, Figure II-14, can be designed to the diameter dictated by payload requirements. The adapter is mounted directly on the instrument unit and is a 45 degree conical frustrum of semimonocoque construction. The adapter includes a structural ring to bear the lateral components of the loads imposed by the payload and to provide clearance for the end frame of the fairing at Station 1704.8. The height of the adapter above the mating plane at Station 1704.8 is shown as 36 inches. This dimension can be varied slightly, if required. The adapter would weigh about 200 lb.

Configuration "A" shown in Figure II-11 utilizes the volume that would be available if mission objectives are such that the LEM is not used. A payload volume of about 3,230 cubic feet and a weight of 7,900 lb could be accommodated in this space. The payload could remain with the S-IVB in orbit, or be ejected after separation of the forward section of the spacecraft/LEM adapter. The LEM adapter panels can be unfolded.

Configuration "B" shown in Figure II-12 is used with the Voyager spacecraft and encompasses a volume of approximately 2,990 cubic feet. These dimensions are approximate and should be used for preliminary layout only. The final dimensions depend on payload configuration and adapter height requirements. The approximate weight of the fairing is 2,500 lb, if made of aluminum honeycomb.

Configuration "C" shown in Figure II-13 is for a modified LEM adapter and encompasses a volume of about 5,000 cubic feet with the approximate dimensions shown. The weight of the fairing is about 3,000 lb.

Prime Payload Attitude Control Systems

Prime payloads may require their own attitude control systems if they are to be separated from the S-IVB stage during orbital coast.

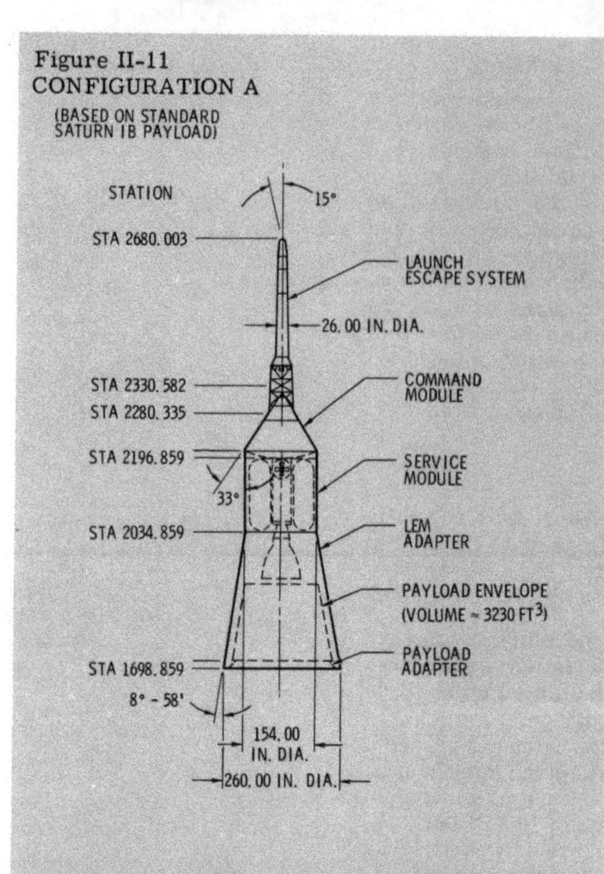

Figure II-11
CONFIGURATION A
(BASED ON STANDARD SATURN IB PAYLOAD)

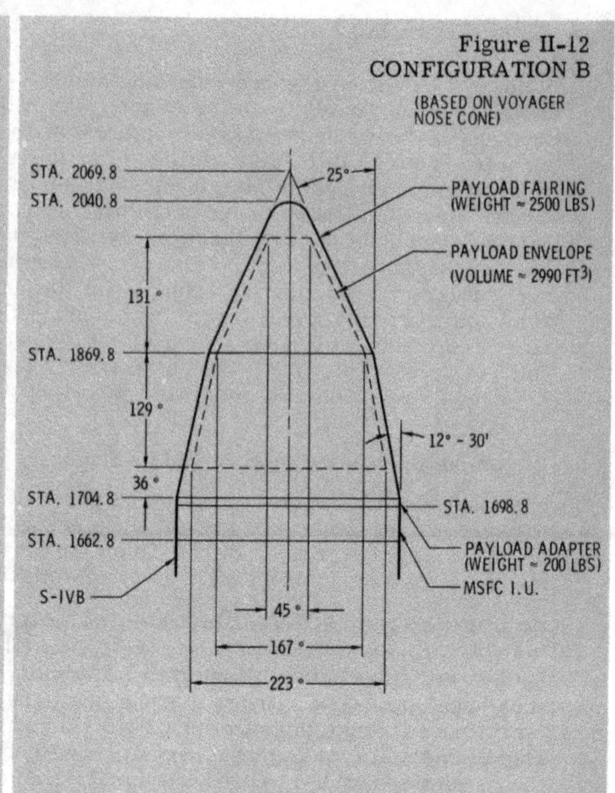

Figure II-12
CONFIGURATION B
(BASED ON VOYAGER NOSE CONE)

* Approximate Dimensions to be Used For Preliminary Layout Only

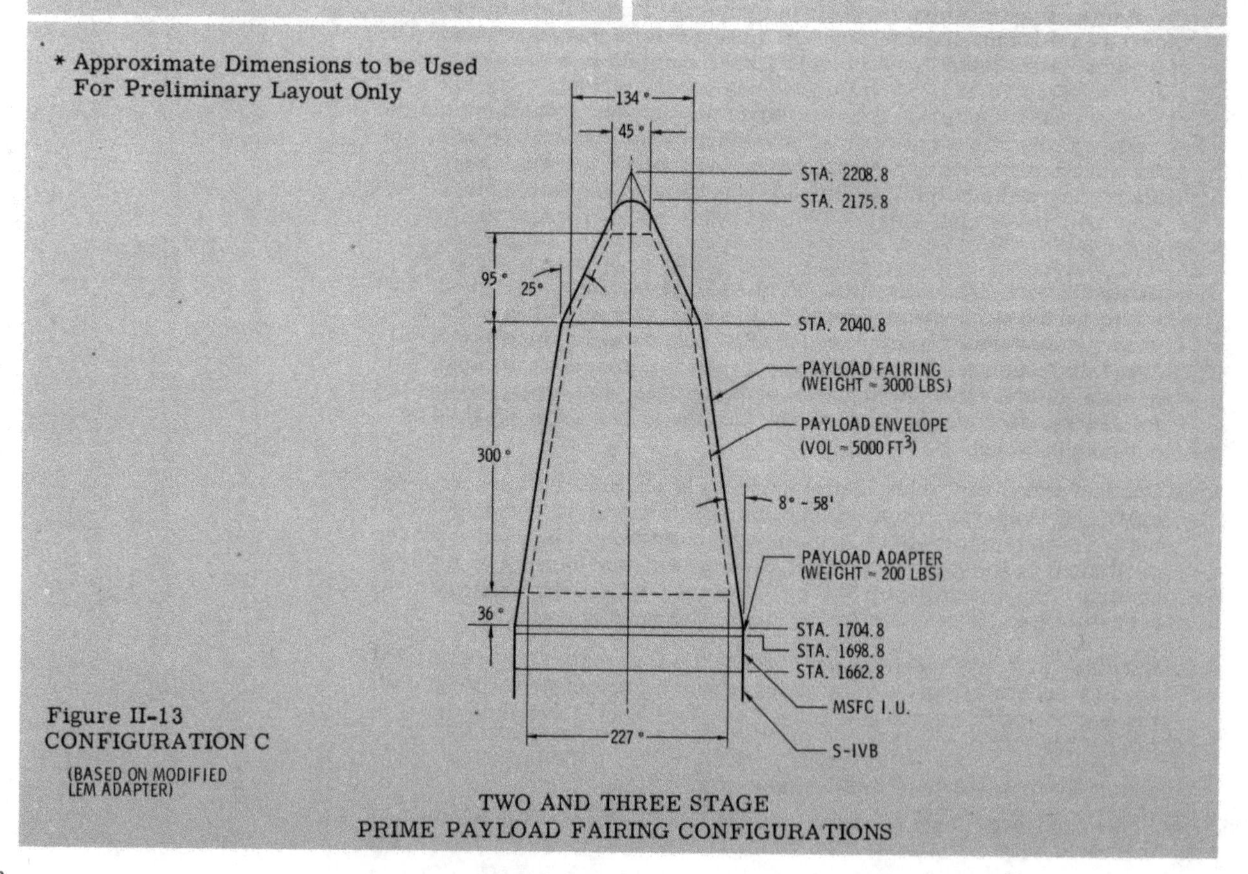

Figure II-13
CONFIGURATION C
(BASED ON MODIFIED LEM ADAPTER)

TWO AND THREE STAGE
PRIME PAYLOAD FAIRING CONFIGURATIONS

Figure II-14
PRIME PAYLOAD ADAPTER

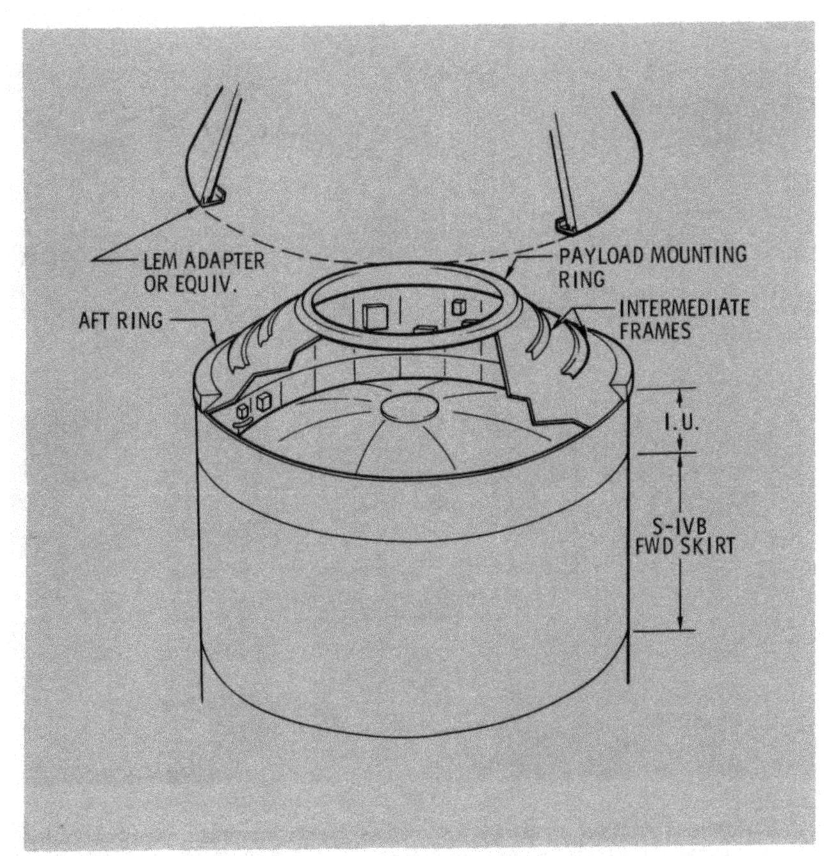

A concept using existing Saturn IB/S-IVB or Saturn V/S-IVB Auxiliary Propulsion System (APS) modules on prime payloads is presented in Figure II-15. The APS modules are presently designed for 4-1/2 and 6-1/2 hours coast, respectively, for the Saturn IB and V missions. Longer coast periods will be achievable if these units are used for payload attitude control.

The APS modules are self-contained propulsion units which require electrical power, vehicle attitude sensors, control circuitry and guidance signals. The guidance and attitude sensing signals are provided by the I.U. The electrical power requirements for either the Saturn V or IB modules are 28 volts at a maximum of 26.5 amp for operating valves and switches. The attitude control band requirements of the payload, moments of inertia, center of gravity, location of the payload, and environmental disturbances dictate the total propellant needed for a given mission. The 150 lb thrust is, perhaps, larger than necessary but there are techniques available for reducing it by about 50% for better propellant economy. Smaller engines from other programs could also be used. However, with the payloads indicated in Figure II-15, for the Saturn V module, control periods in excess of 6-1/2 hours with a deadband of $\pm 1^{o}$ in all three axes are possible. Reducing the control accuracy requirements extends the operating duration to 24 hours or more. Detailed descriptions of the APS modules are presented in Section II.E.

C. Payload Thermal Environment and Control

Accurate determination of the payload temperature control requirement demands realistic thermodynamic models. The analytical tool Douglas uses to obtain these is a series of heat transfer computer programs (both one-and three-dimensional) which indicates the temperature history that can be expected at any point in the vehicle for a multitude of thermodynamic environments. Should the temperature of the volume be critical for a particular experiment, a thermal protection system can then be designed.

Figure II-15

PRIME PAYLOAD USING S-IVB AUXILIARY PROPULSION SYSTEM

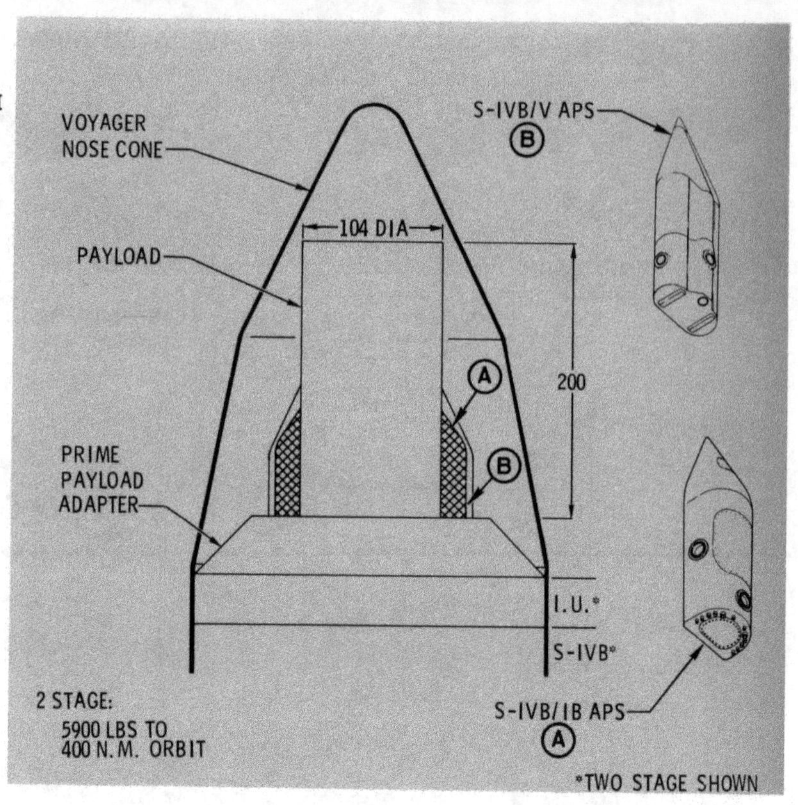

The thermal control systems for the electronic equipment mounted in the forward skirt and that mounted in the aft skirt are different. The equipment mounted on the forward skirt is conditioned actively and that mounted aft is conditioned passively.

In the active system, electronic components are mounted on thermal conditioned panels (cold plates, Figure II-9) which transfer heat to a coolant (60% methanol and 40% water) flowing through the plate. For the present S-IVB and I. U. flight plans, the coolant will enter the cold plates at 60°F maximum and leave at 70°F maximum. A coolant flow rate of approximately 0.5 gpm per panel is used at presant. Since each part (i.e., resistors, transistors, etc.) in a component operates at a different temperature and since there is no gas medium to provide convective heat transfer passing through the components while in space, it is difficult to define average operating temperatures of the components. Units generating high heat loads should not be mounted close together since the coolant may not be capable of removing the required heat and excessive part temperatures will result. The total allowable heat load per panel is 500 watts. Douglas can provide assistance in solving these problems.

The experimenter must consider the weight to be mounted on the cold plates. The mounting methods for the cold plates and the vibration levels predicted during launch allow 150 lb of equipment on each plate. Concentrated loads should be avoided. Experiments must be designed so that they can be mounted on the plates without interfering with coolant channels.

No cooling is available from the end of the prelaunch phase until approximately 130 sec after lift-off. A pre-launch purge gas system, utilizing air and gaseous nitrogen, provides the forward skirt area with a warming medium. It operates only up to the time of launch and provides no thermal control after that time. This system protects the electronic components and reduces oxygen present to 4% by volume. The total flow rate in the forward area is about 275 lb/min. The purge gas surrounding the components located in the I. U. and S-IVB forward skirt will be at a temperature of 35°F to 75°F.

Electronic equipment mounted in the aft skirt area is attached to fiberglass panels. No fluid thermo-conditioning system is used. Temperature is controlled through proper surface finish of each electronic package and by providing conduction paths and insulation. Appropriate coatings are added when a special heating or cooling problem is revealed by calculation or test.

As now designed, the fiberglass panels in the aft skirt area can support 100 lb each of electronic packages.

A separate pre-launch purge gas system maintains the equipment mounted on the aft skirt at a temperature of 20°F to 70°F during pre-launch procedures. Dry air at a flow of 300 lb. per minute to the S-IVB stage is provided from a ground source. Gaseous nitrogen purge of approximately the same flow rate is initiated about 30 minutes before LH_2 loading. During flight, heat is radiated to space and to local sinks such as the LOX tank. If possible, high heat dissipating components or temperature sensitive components should be mounted forward on the cold plates.

If the above systems do not meet the needs of an experiment, modifications can be made to the thermal conditioning system or the purge gas system. Example of such changes are:

(1) Coolant flow rate in the thermal conditioning system can be changed to control the temperature of the experiment equipment.

(2) A space radiator could be installed to cool electronic equipment for long periods of time.

(3) Insulation and thermal control coatings may be engineered.

(4) Mounting procedures and requirements can be altered to vary heat conduction paths.

(5) Flow rate and temperature of the purge gas system could be varied.

(6) Purge gas could also be ducted directly to the experiment equipment.

D. Payload Acoustics and Vibration Environment

Acoustic and vibration phenomena have a similar time-history during a flight. A time-history of the former is shown in Figure II-16. The acoustic noise level inside the vehicle at three auxiliary and prime payload locations are shown in the figure. At lift-off, the exhaust of the first stage engines' generates high frequency noise in the heavily sheared mixing region close to the nozzles and lower frequency noise from the fully turbulent cores of the exhaust jets. This is transmitted through the air to the spacecraft and vehicle and induces vibration.

Following lift-off, the acoustic noise decreases as the exhaust pattern straightens out and as the distance between the vehicle and the ground reflecting surface increases. A further reduction occurs as the vehicle reaches supersonic speeds because the sound generated aft of the vehicle is unable to catch-up with the vehicle. In the meantime, turbulent pressure fluctuations in the aerodynamic boundary layer are intensifying as free stream dynamic pressure increases. The maximum noise from this source occurs at the time of maximum dynamic pressure or shortly thereafter and it falls off as the dynamic pressure decreases with air density. The remaining excitation is structurally transmitted from engines, pumps, etc. These are of lower intensity and remain relatively constant until engine cut-off. Brief periods of vibration occur during retro rocket and ullage rocket firings and stage separation.

To provide design criteria for payloads, design specifications have been developed which cover all of these environments. Figure II-17 is a broad-band acoustic specification for three payload locations, and represents an acoustic environment to which an item may be designed and ground tested to ensure satisfactory operation during an actual flight. All frequencies are assumed to be excited at the same time and at the appropriate level in each octave band. Figure II-18 is a broad-band random vibration specification for the same purpose. The duration of these qualification tests is longer than the duration of the significant environment for an actual flight to allow for exposure during static firings and to increase the reliability of the items. Figure II-19 is a sinusoidal vibration specification. The purpose of the sinusoidal sweep test requirement is to provide assurance that the item has adequate strength for transitory or unsteady phenomena that could occur in a flight. There is also a shock specification for each location but it is not included in this brief discussion.

E. S-IVB Stage Subsystem Information

There are at least four major subsystems of the S-IVB stage that may influence or be of benefit to an auxiliary payload. They include the auxiliary propulsion, electrical power, thermal conditioning, and data acquisition systems.

1. Auxiliary Propulsion System

The Auxiliary Propulsion System (APS) modules for the S-IVB stage have two basic configurations as indicated by Figure II-20. The two are necessary to meet the mission requirements of the Saturn V vehicle and the Saturn IB vehicle. The major differences between the two are in propellant capacity and ullage capability.

The Saturn IB/S-IVB APS is sized to provide roll control during powered flight and three axis attitude control in pitch, roll and yaw during a 4-1/2 hour earth orbital coast. A mock-up of the Saturn IB/S-IVB module is shown in Figure II-21. The Saturn V/S-IVB APS provides these same functions and, in addition, provides three-axis attitude control during a two-hour translunar coast period. A 70-pound thrust engine is also contained in each Saturn V module to provide propellant settling for continuous vent initiation and main engine restart.

The attitude control system is a pulse-modulated on-off system. The system is based on the minimum impulse capability of the 150-lb thrust engine, which has a minimum impulse bit capability of 7.5 lb-sec with an electrical input pulse width of approximately 65 milliseconds. The attitude control system is designed to operate with an attitude dead zone of ±1 degree in all axes. The undisturbed limit cycle rates of the Saturn IB/S-IVB with payload in a 100 n.mi. orbit are 0.03 deg/sec in roll and 0.005 deg/sec in pitch and yaw.

The auxiliary propulsion system, which provides these stage requirements, is a completely self-contained modular propulsion sub-system. The modules require electrical power and command signals to provide the necessary stage functions. They are mounted on the aft skirt 180° apart. The equipment for loading propellants to the modules is a semi-automatic system, with individual umbilical connectors in each module.

Each module contains three 150-lb thrust ablatively-cooled liquid bi-propellant hypergolic attitude-control engines, a positive expulsion propellant feed system for zero gravity operations and a helium pressurization system. Each Saturn IB APS module contains 23.3 lb of MMH (Monomethylhydrazine) fuel and 37.7 lb of N_2O_4 (nitrogen tetroxide) oxidizer. The nominal oxidizer to fuel mixture ratio is 1.65.

Figure II-16

ACOUSTIC NOISE TIME-HISTORY

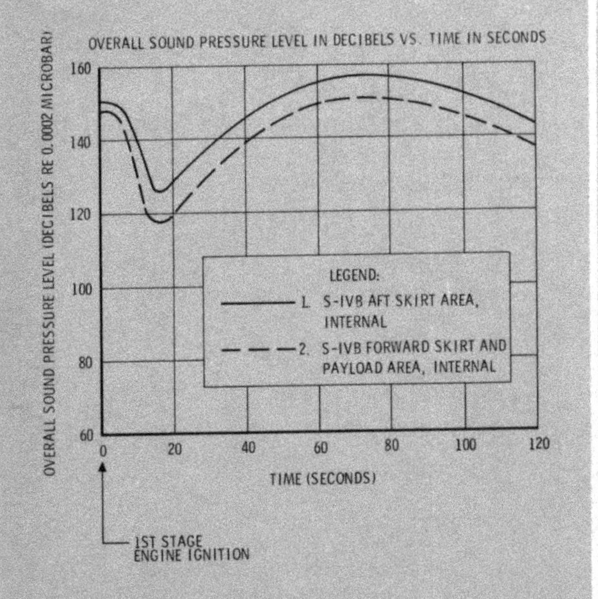

Figure II-17

DESIGN SPECIFICATION FOR ACOUSTIC NOISE

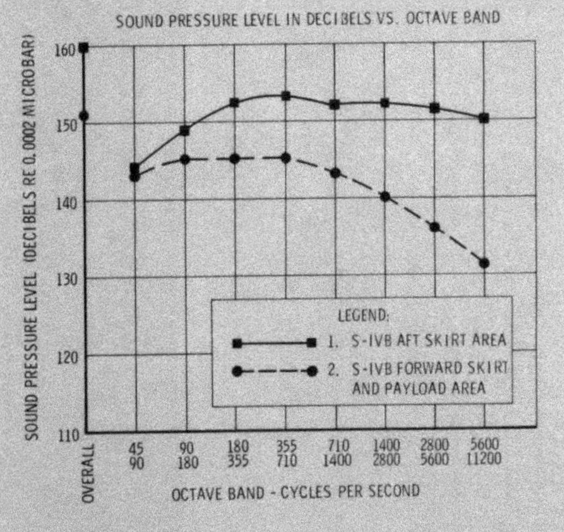

NOTE:
1. THE TIME DURATION IS ASSUMED TO BE EIGHTEEN MINUTES.
2. THE DIFFUSED SOUND FIELD OF RANDOM NOISE IS ASSUMED TO HAVE A GAUSSIAN AMPLITUDE DISTRIBUTION.

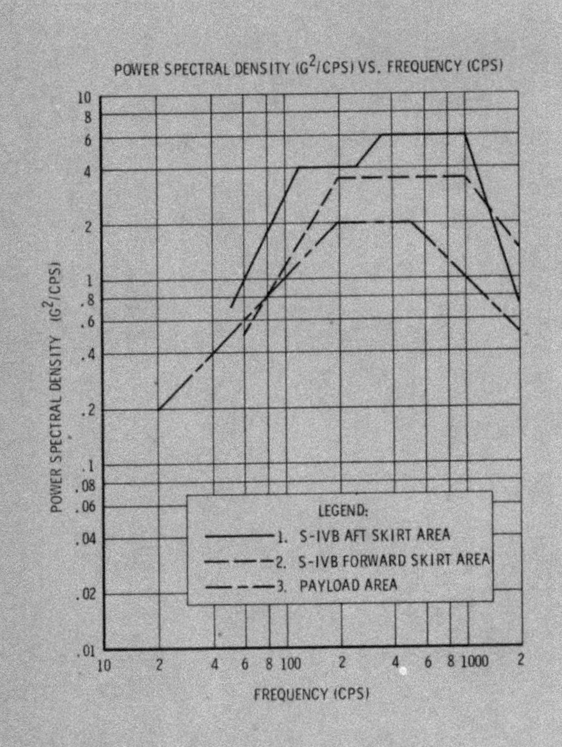

NOTE:
1. AMPLITUDE DISTRIBUTION IS ASSUMED GAUSSIAN
2. DURATION IS ASSUMED TO BE TWELVE MINUTES FOR EACH OF THREE MUTUALLY PERPENDICULAR DIRECTIONS

Figure II-18

DESIGN SPECIFICATION FOR RANDOM VIBRATION

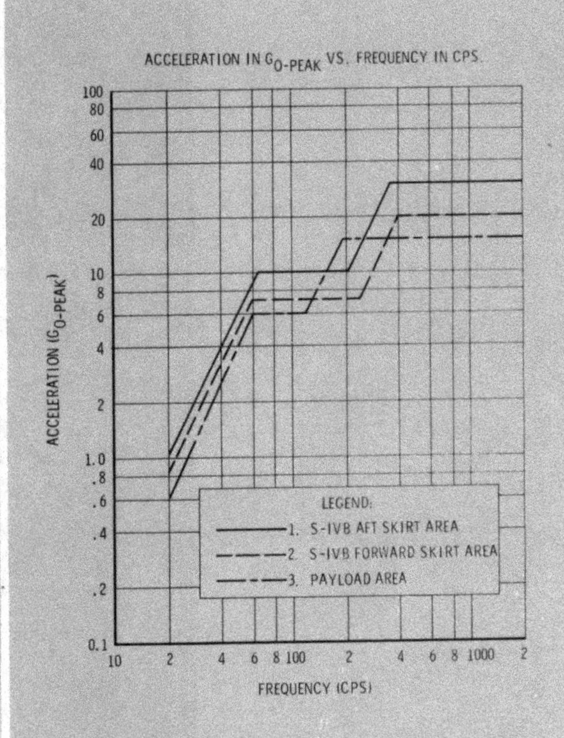

NOTE:
1. THE VIBRATION INPUT IS ASSUMED TO BE APPLIED IN EACH OF THREE MUTUALLY PERPENDICULAR DIRECTIONS.
2. THE LOGARITHMIC SWEEP RATE IS ASSUMED TO BE ONE OCTAVE PER MINUTE OVER THE FREQUENCY RANGE FROM 20 TO 2000 AND BACK TO 20 CPS.

Figure II-19

DESIGN SPECIFICATION FOR SINUSOIDAL VIBRATION

Figure II-20
S-IVB AUXILIARY
PROPULSION SYSTEM

SATURN IB

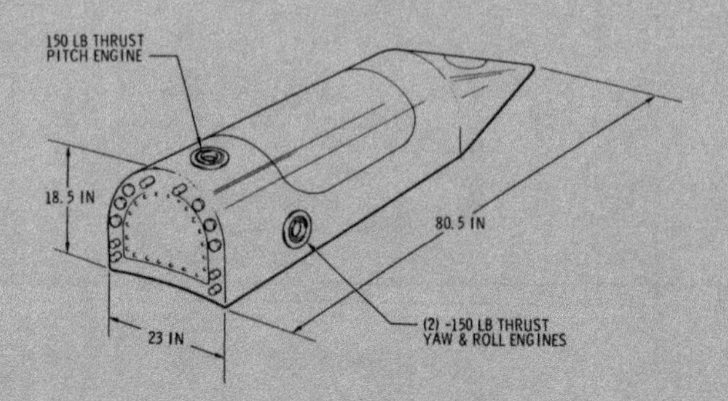

SATURN V

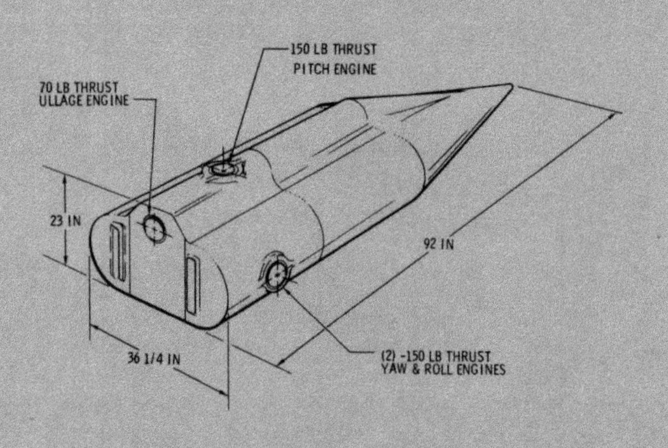

	SATURN V	SATURN IB
TOTAL MODULE DRY-WEIGHT	506 LBS	422 LBS
TOTAL WEIGHT OF LOADED MODULE	818 LBS	483 LBS
TOTAL PROPELLANT CAPACITY	312 LBS	61 LBS
MIN. TOTAL IMPULSE (0.065 SEC/PULSE)	65,000 LB-SEC	14,000 LB-SEC
NOMINAL TOTAL IMPULSE	70,000 LB-SEC	15,000 LB-SEC
MAX. TOTAL IMPULSE AVAILABLE	75,000 LB-SEC	16,200 LB-SEC

Figure II-21

SATURN IB/S-IVB AUXILIARY PROPULSION SYSTEM MODULE (MOCK-UP)

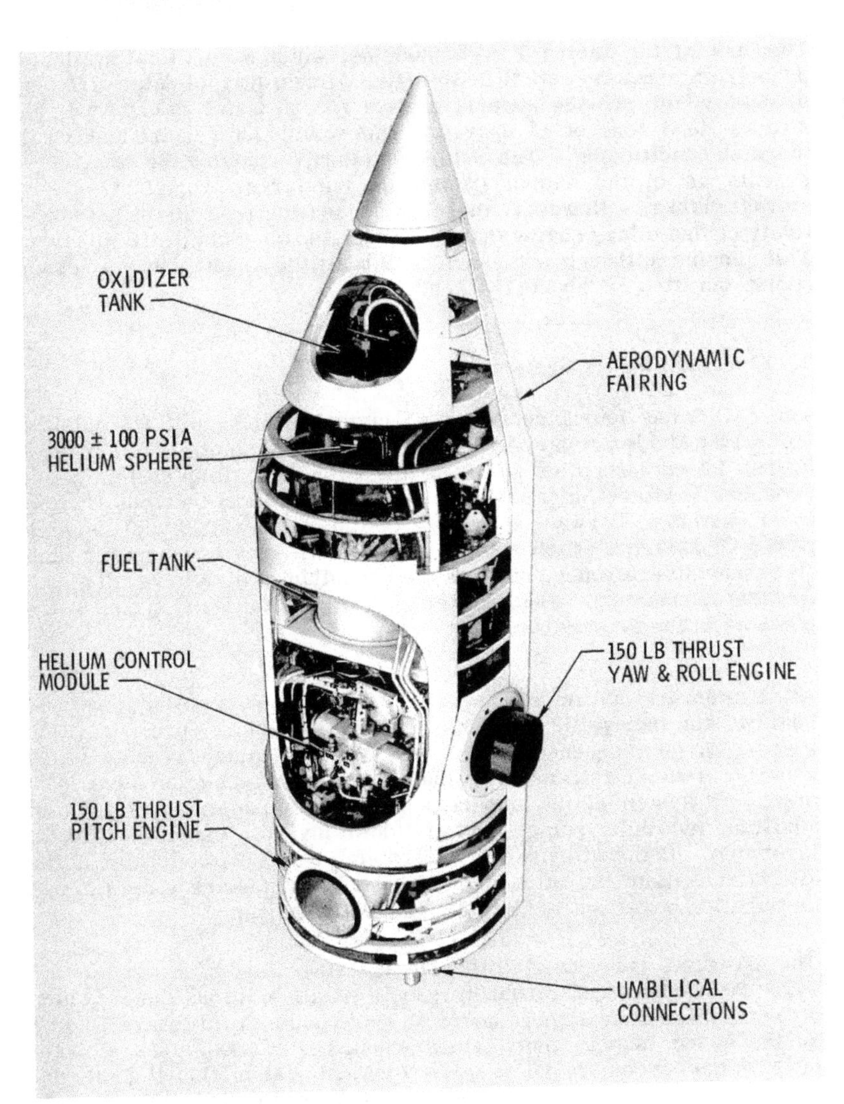

The total firing time for the engines is 7 minutes for steady state operations. Pulse operation at a pulse frequency of up to 10 pulses per second is possible. Testing has demonstrated a pulse mode capability of over 20 minutes accumulated burn time.

Many attitude and maneuvering exercises other than those now required could be performed. The mission time could be extended by increasing the attitude dead zone to reduce propellant consumption. Greater attitude control accuracy is also possible, but at the expense of mission time. Propellant conditioning to prevent freezing may be required for extended durations in space. This could include propellant additives, to lower propellant freeze points, passive insulation or active thermal protection.

If more total impulse is required for an experiment or mission than can be provided by the two standard Saturn-IB modules, the aft skirt of the S-IVB could be strengthened to allow the use of four Saturn IB modules, or two Saturn V modules. Only two of the Saturn IB modules would be operated at any one time. Control signal switching circuitry would have to be added to reverse the pitch-yaw planes when the first two modules are depleted.

The use of the Saturn V APS modules, which have a total available propellant capacity of 312 lb (five times that of Saturn IB) per module could provide attitude control for 30 hours in orbit with an attitude dead zone of ±1 degree. This would also require additional thermal conditioning. The Saturn V module contains the same components as in the Saturn IB module but larger propellant and gas storage tanks. However, there is an additional 70-lb thrust ablatively-cooled ullage engine that can be deleted for a Saturn IB mission. This engine although not required for attitude control has a design firing duration of about 454 sec.

2. Electrical Power System

The S-IVB has four independent electrical systems with 56- and 28-volt silver-oxide primary batteries. Forward system #1 (270 ampere hours, 28 vdc) supplies power to the data acquisition system which produces low-level, high-frequency signals that must be isolated from other systems. Forward system #2 (4 ampere hours, 28 vdc) supplies power to systems which cannot tolerate switching transients or high frequency interference, such as the propellant utilization system and inverter-converter. Both batteries for the forward systems are mounted in the forward skirt.

Aft system #1 (50 ampere hours, 28 vdc) supplies power to valves, heaters and relays in the main propulsion engine, pressurization system, stage sequencer, APS modules and ullage rockets which generate switching transients, that must be isolated from other systems. Aft system #2 (25 ampere hours, 56 vdc) supplies power for an auxiliary hydraulic pump, LOX chilldown inverter and LH_2 chilldown inverters. Both batteries for aft system #1 and #2 are located in the aft skirt. Both the aft and the forward systems are wired through distribution boxes located in their respective areas.

The batteries are sized to handle the stage load requirements for 4-1/2 hours but could probably handle small additional loads, which do not exceed four ampere hours at a maximum drain rate of 10 amp on the 56 vdc supply and 70 ampere hours, at a maximum drain rate of 15 amp on the 28 vdc supply (forward system #1). If additional power is required for the planned 4-1/2 hours or for longer periods, additional batteries or larger Saturn V batteries could perhaps be used for as much as 72 hours which is the shelf life of the batteries. If power is required for even longer, other batteries or fuel cells could be used.

3. Data Acquisition System

The present Saturn IB R&D vehicles have five telemetry systems; one single sideband/frequency modulation (FM) system, one pulse code modulation (PCM) /FM system, and three FM/FM systems. One channel of each FM/FM system will be used for sampled pulse amplitude modulation data. Pertinent data on these systems are shown in Table II-I.

Vehicle SA-205, to be delivered in mid-1966 and all subsequent vehicles will have only one telemetry system (PCM/FM). The capability of this operational telemetry system is:

8 channels at 120 samples/second
460 channels at 12 samples/second
190 bi-level using remote digital sub-multiplexer
658 total measurement capability

or

54 channels at 120 samples/second
190 bi-level using remote digital sub-multiplexer
244 total measurement capability

TABLE II-I

SATURN IB R & D TELEMETRY SYSTEMS

	T/M System	Frequency (MC/S)	Prime Channels	Prime Sampling Rate (Per sec)
1.	SS/FM	226.2	15	Continuous
2.	PCM/FM	232.9	0-100 Bi-level +	120
			Parallel Acceptance of 3 PAM Multiplexers at	120, 40
3.	FM/FM	246.3	15	Continuous
	PAM/FM/FM		30	120
4.	FM/FM	253.8	15	Continuous
	PAM/FM/FM		30	120
5.	FM/FM	258.5	15	Continuous
	PAM/FM/FM		30	120

		Total Measurement Capability	
1.	SS/FM	15 prime channels possible to sub-multiplex by 5 =	75
2.	PCM/FM	100 Bi-level channels + 30 prime channels on checkout multiplexer: 3 prime channels for frame sync & calibration; 23 prime channels possible to sub-multiplex by 10 =	334
3 - 5	FM/FM	15 prime channels possible to sub-multiplex by 3 =	45
	PAM/FM/FM	30 prime channels per multiplexer: 3 prime channels for frame sync & calibration; 23 prime channels possible to sub-multiplex by 10 = 234 x 3	702
			1,156

The 658 or the 244 measurement capability is based on the utilization of two multiplexers. If increased to four multiplexers, the capability becomes:

230 channels at 12 samples/second
4 channels at 120 samples/second
690 channels at 4 samples/second
12 channels at 40 samples/second
190 bi-level using remote digital sub-multiplexer
1126 total measurement capability

Additional combinations of sampling rates are obtainable. It is estimated that about 50 channels would be available for auxiliary payloads. The exact number of telemetry channels available can only be determined after the vehicle is selected because instrumentation varies from one vehicle to another. Once a payload application is established and scheduled, and bandwidth, accuracy, etc. are known, a determination of available channels can be made.

The operational vehicles (all vehicles after SA-204) also have provisions for mounting one complete set of modified R&D FM/FM system in kit form. Its capability is: 18 channels of continuous data, or with sub-multiplexing, 32, 64 or 80 channels depending upon the sampling rate.

If the payload originator has a reason for furnishing part or all of his own Data Acquisition System, this could be arranged.

F. Orbital and Deep Space Tracking, Data and Control Stations

Requests for payload data must be integrated into the overall mission plan and approved by the appropriate NASA office. Orbital or space tracking and control functions required by a payload after separation from the Saturn must also be specifically approved.

G. Launch Support Facilities

The Saturn IB uses launch Complex 34 and 37 of the Eastern Test Range. Complex 34 has one pad while Complex 37 has two (Pad A and B). At present only Complex 34 and Complex 37B will be used. Figures II-22 and II-23 show a photograph of Complex 37B with Saturn I on the pad and an illustration of the layout of Complex 37. During the last three Saturn I flights, which will be launched from Complex 37B, Complex 34 will be modified for Saturn IB. Following the Saturn I flights, Complex 37B will be modified for Saturn IB launches.

Complex 37 launch pad is 47 feet square and 35 feet tall, with a 33-foot diameter opening in the center to channel the engine exhaust gases onto the deflector below. The pad is also used as a base upon which to mount support equipment, such as the tower containing electrical, pneumatic and propellant loading umbilicals.

The umbilical tower, which stands alongside the launch pad, is 268 feet high and designed so that as much as 52 feet of structure could be added should additional height be required. It supports umbilical swing arms which carry the electrical, environment-control, pneumatic, and propellant vehicle servicing lines during prelaunch preparation and launch. An automatic ground control station (AGCS) is located partly in the base of the umbilical tower and partly below the pad surface under the tower base. It houses the equipment and serves as a terminal and distribution point for measuring and checkout equipment, electrical power control circuitry for propellant and high-pressure gas system, pneumatic control equipment, and piping systems connecting the vehicle and its ground support equipment. A generator room there provides AC and DC power. A cable tunnel under the pad connects the AGCS room to the cableway extending to the launch control center.

The launch service structure is 300 feet high. It is designed so that 30 feet of structure can be added if needed. A derrick mounted atop of the launch service structure lifts all stages and spacecraft during the erection of the launch vehicle. The service structure is equipped with a clamshell "silo" to enclose the vehicle on the pad for hurricane protection. Adjustable work platforms give access to all parts of the vehicle during launch preparation.

Complex 37 is equipped with a central high-pressure gas system to supply gaseous nitrogen and helium. Liquid nitrogen is stored in a 35,000-gallon insulated tank. A gaseous-nitrogen storage tank is supplied by vaporizers in the liquid-nitrogen storage tank. The gaseous helium storage supply is charged by compressors from supplies of helium brought in at lower pressures.

The blockhouse, housing the Launch Control Center (LCC), is a circular dome of reinforced concrete. From it can be controlled all prelaunch and launch activities. It can checkout the complete vehicle

Figure II-22
[illegible] ON PAD 37B KSC

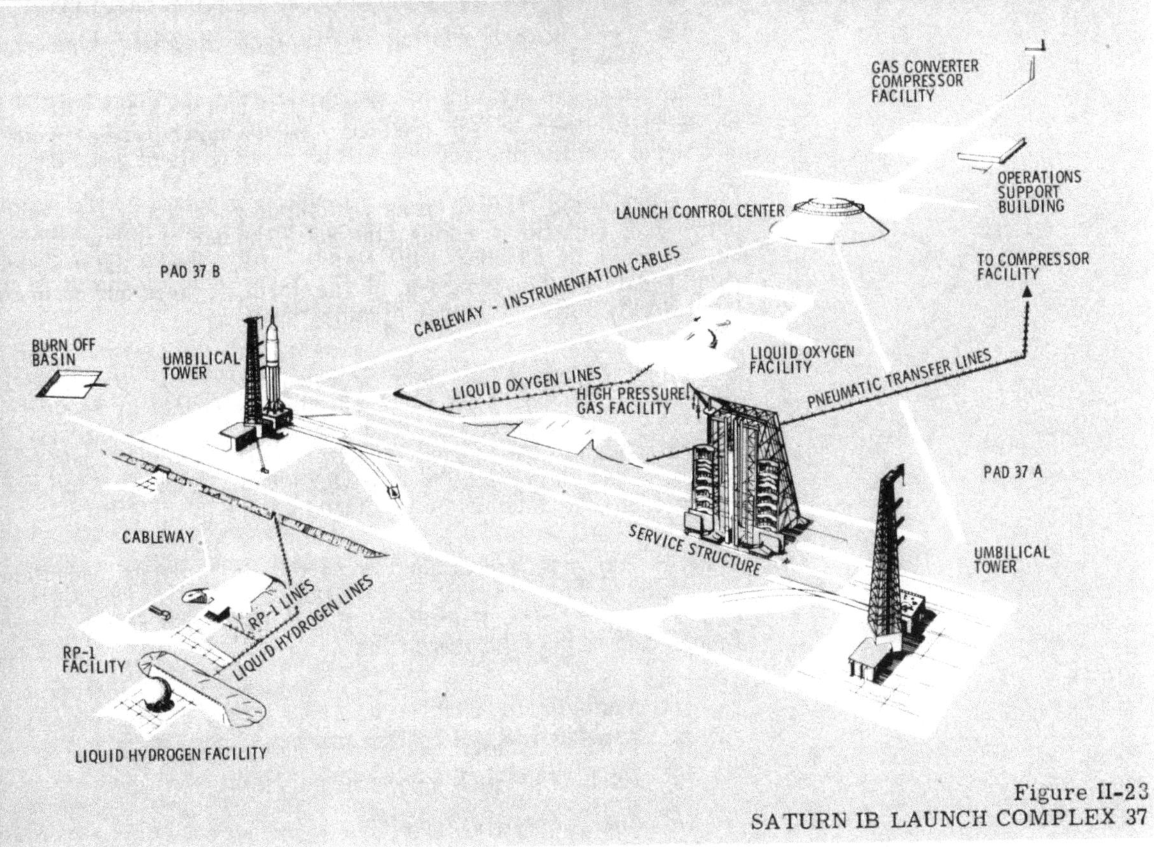

Figure II-23
SATURN IB LAUNCH COMPLEX 37

automatically with an RCA 110 A computer system. The Launch Control Center is approximately 1150 feet from the launch pad and has visual contact via periscopes and by television.

In addition to the above facilities, the Eastern Test Range has pulse radar systems, infrared systems, metric optics (ballistic cameras, cinetheodolite, fixed cameras), documentary photography, missile impact location system (underwater sound detection and location) and telemetry (FM/FM, PAM/FM/FM, PDM/FM, Inflight Television, PCM/FM) for tracking and data collection.

The Douglas support facilities for experiment payloads mainly consist of electrical, pneumatic and mechanical prelaunch checkout and handling equipment. The S-IVB stage checkout and handling equipment could possibly be modified to handle checkout functions and handling requirements of an auxiliary or prime payload.

Space can probably be provided nearby for payload originators to park instrumented trailers of their own for remote radio-line checkout if available facilities are not adequate.

H. Data Reduction and Evaluation

Prime payloads (mounted above the S-IVB stage and the I.U.) generally will have self-contained telemetry systems whereas, auxiliary payloads could use available S-IVB or I.U. Telemetry systems. The following information is needed to insure proper use of vehicle Telemetry:

(1) Type of measurement (pressure, temperature, signal, vibration, strain, etc.)

(2) Range and accuracy of measurement needed

(3) Type of monitoring (continuous, sampling, real time)

(4) Type of presentation (punch tape, magnetic tape or strip chart)

The interconnect between the test article and the stage will be made by standard flight proven methods. Where appropriate, transducers and signal conditioning devices will be existing flight qualified items.

Data transmitted from orbiting vehicles is received by the worldwide network of satellite tracking stations. The use of this network must be arranged in advance with NASA. Procedures exist which will permit data to be recorded in the form of tapes and strip charts from NASA, USAF, and other organizations.

A limited amount of data may be received prior to lift-off via stage umbilicals and would be recorded on strip charts or sequence recorders. All data telemetered from the vehicle, is received and recorded during pre-launch, launch, and orbit. Vehicle data may aid reduction and interpretation of payload data. Quick look and in some cases, real time data can be provided by the ETR Facilities.

Douglas can give experimenters partial or complete reduction and evaluation of data in a final report. Douglas has data reduction facilities at the Douglas Huntington Beach Data Laboratory which can reduce data to the following forms:

(1) Analog strip charts

(2) Tabulated digital readouts in engineering units

(3) Machine printed, scaled analog plots

(4) Analog oscillograph plots

The Huntington Beach Computer Facility is equipped with IBM 7094 computers for lengthy, iterative, computation processes.

I. Checklist for Auxiliary Payload/Vehicle Interface Requirements

The first important step in planning to propose an auxiliary payload for Saturn IB is to document information on the physical and operating characteristics of the payload along with the required launch vehicle accommodations and ground support. With such information, it will be possible for you to discuss with us the various aspects of Saturn IB flight accommodations. The information which we provide to you regarding the payload/launch vehicle interface can beneficially augment the proposal which you submit to the appropriate NASA organization for review. A check list is given below of the items of information typically required to properly consider and define Launch Vehicle accommodations for your payload. Information on the experiment submittal process and associated vehicle data can also be obtained from cognizant NASA Agencies.

A. General Information Required

Experiment title

Proposal Originator

Purpose and application of experiment

Relationship to Apollo or other national goals

Description of experimental procedures

Present status of experimental equipment

Scope of budget or available funding

B. Experiment Mission Requirements

Orbital Altitude, circular, elliptical (apogee-perigee)

Suborbital or orbital flight durations, minimum, maximum

Desired launch azimuth

Desired launch inclination

Desired date of launch (year)

Astronauts' time required, pre-flight, inflight, post-flight

C. Experimental Equipment Capability or Requirements

Envelope description or volume requirements

Weight

Environmental Limitations or Capability

(1) temperature

(2) acoustics

(3) vibration

(4) shock

(5) acceleration

(6) humidity and free moisture

(7) atmosphere and pressure

(8) sand and dust

(9) meteoroids

(10) fungus

(11) salt spray

(12) ozone

(13) hazardous gases

(14) particle radiation

(15) electromagnetic radiation

(16) electromagnetic compatibility

(17) explosion

(18) sterilization requirements

(19) special environmental control

Electrical Power Loads; voltage, current, duration, AC-DC

(1) steady state

(2) intermittent

(3) peak

(4) desired interface locations

Vehicle gas requirements; flowrates, pressures, temperatures

(1) helium

(2) nitrogen

(3) oxygen

(4) hydrogen

(5) others

Jettison Requirements

Special Attitude Control Requirements

(1) stabilization control — deadband

(2) rates of angular acceleration and velocity in pitch, yaw and roll

Schedule Information

Range Safety Requirements

D. Instrumentation Requirements

Type and Numbers: Pressure, temperature, signal, vibration, strain, special

Range and Accuracies

Type of Monitoring

(1) continuous

(2) sampling

(3) real time

Duration or Time period of monitoring

Interface

(1) transducer part of experimental package

(2) transducer part of stage contractor responsibility

(3) location

E. Final Data

Raw data desired

Reduced data desired

Evaluated data desired

Final data package, reports, tapes, graphs, etc.

F. Shroud Design for Prime Payloads

Configuration A, B, C, or special (Figures II-11, 12 & 13)

G. Suggested Mounting Location

H. Ground Support Equipment (Location and Type)

Electrical checkout

Pneumatic

Mechanical

Handling

Servicers

I. Tracking, Data Acquisition and Command

J. Facilities

Facilities needed by payload originator at Douglas Space Systems Center or Kennedy Space Center.

K. Special

Any special requirements that affect the integration of the payload with the launch vehicle.

If your experiment is a prime payload for Saturn IB, this same information will be required by the cognizant NASA agency.

If you desire help in integrating your experiment with the Saturn IB vehicle, please forward your request to the address shown in the foreword of this guide.

Figure III-1
OPERATIONAL SATURN IB CONFIGURATIONS

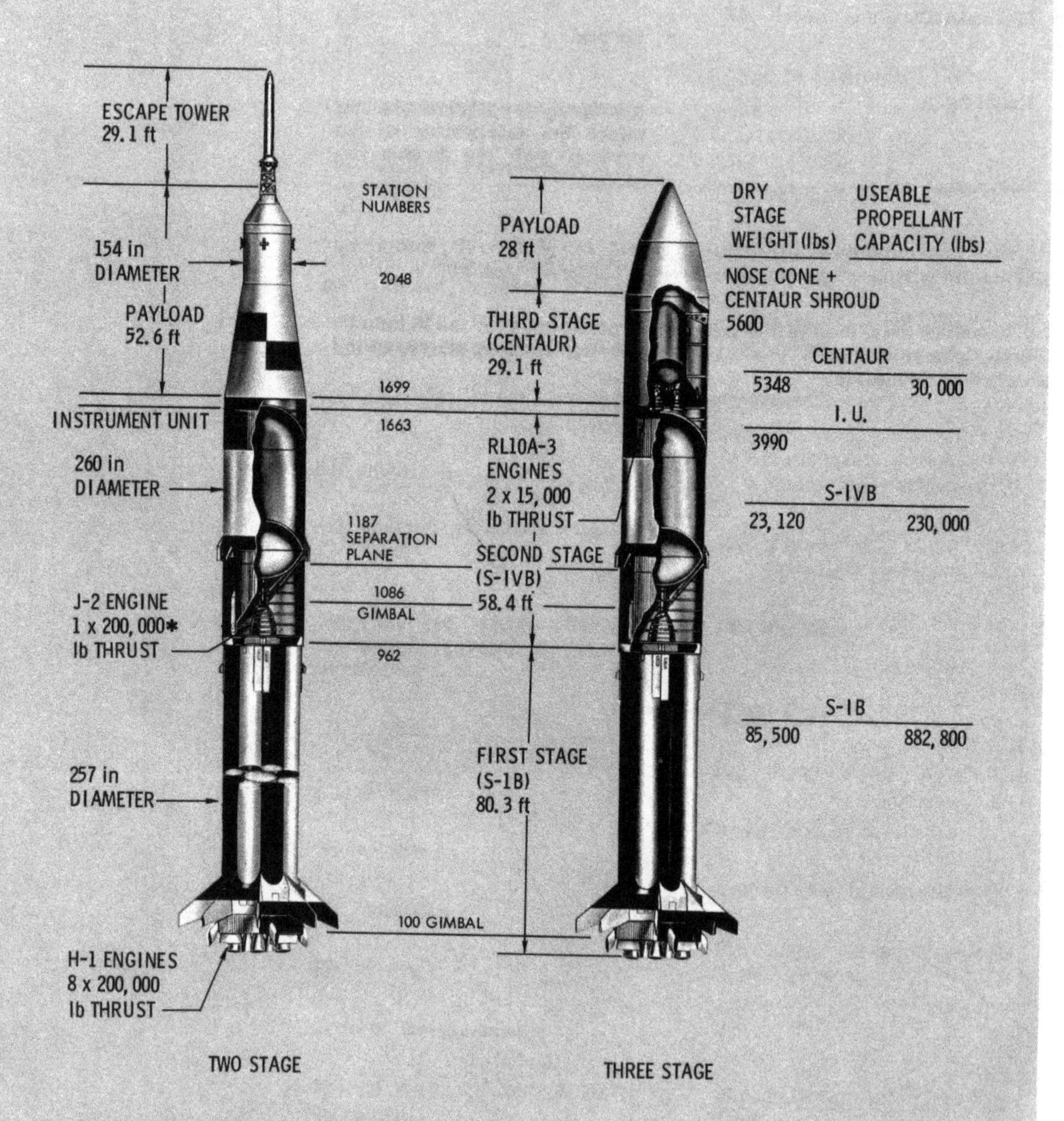

✱ 205,000 lb Thrust After
Vehicle SA-205 or 206

III. SATURN IB CONFIGURATIONS

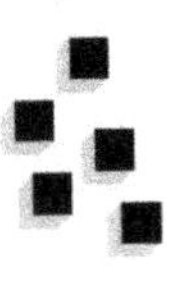

The two-stage configuration of the Saturn IB vehicle is in production and a three-stage version is planned. The first two stages of both are the S-IB, built by Chrysler Corporation Space Division of New Orleans, La., and the S-IVB, built by Douglas Aircraft Company Missile and Space Systems Division of Huntington Beach, California. An Instrument Unit built by International Business Machines Corporation (IBM) of Owego, New York, is mounted above the S-IVB and houses the guidance, control and non-stage oriented flight instrumentation systems.

The third stage, when used, is the Centaur. It is built by the General Dynamics Corporation of San Diego, California. Official definitions of the stages are contained in contractual specifications governing each stage contractor. The two operational configurations of the Saturn IB are presented in Figure III-1.

The Saturn IB vehicles which will fly prior to Saturn V, represent an extension of Saturn I as depicted in Figure III-2. The two-stage Saturn IB vehicle, less payload, weighs approximately 1,250,000 lb at lift-off and can place up to 35,000 lb into a 100 nautical mile earth orbit. Its first-stage engines generate a total of 1,600,000 lb of sea level thrust. The three-stage Saturn IB vehicle, less payload, weighs about 1,282,000 lb at lift-off, can place 9800 lb in synchronous orbit (with no plane change), and can accelerate 12,300 lb to escape velocity.

The Saturn vehicles stand alone in their payload class and have practically unlimited applications to both manned and unmanned earth orbital or interplanetary missions. A single large payload may be the primary purpose for launching the Saturn IB vehicle, but many small auxiliary scientific or engineering payloads can be carried into space economically aboard the S-IVB stage.

A. Two Stage Saturn IB

1. First Stage (S-IB)

The S-IB first stage of the Saturn IB vehicle is an improved version of the Saturn I first stage (Figures III-3 and -4). This stage is manufactured by the Chrysler Corporation Space Division, Michoud Operations near New Orleans, Louisiana. The S-IB stage uses nine separate tanks, a cluster of eight 70-inch diameter tanks (four fuel and four oxidizer) surrounding a 105-inch diameter center oxidizer tank.

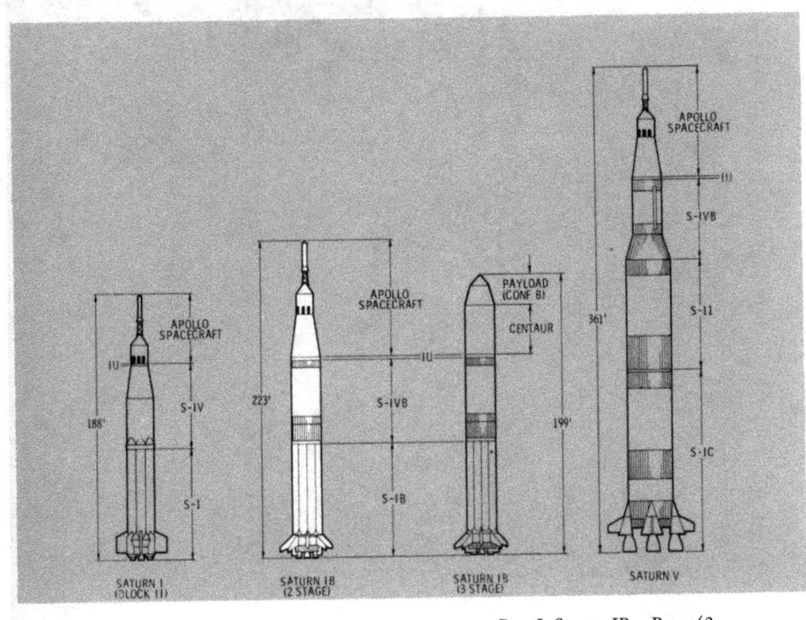

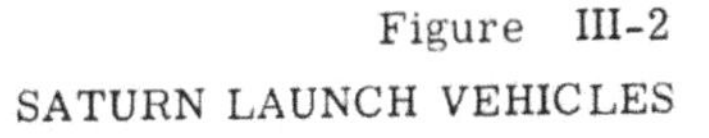
Figure III-2

SATURN LAUNCH VEHICLES

Figure III-3
SATURN IB/S-IB STAGE

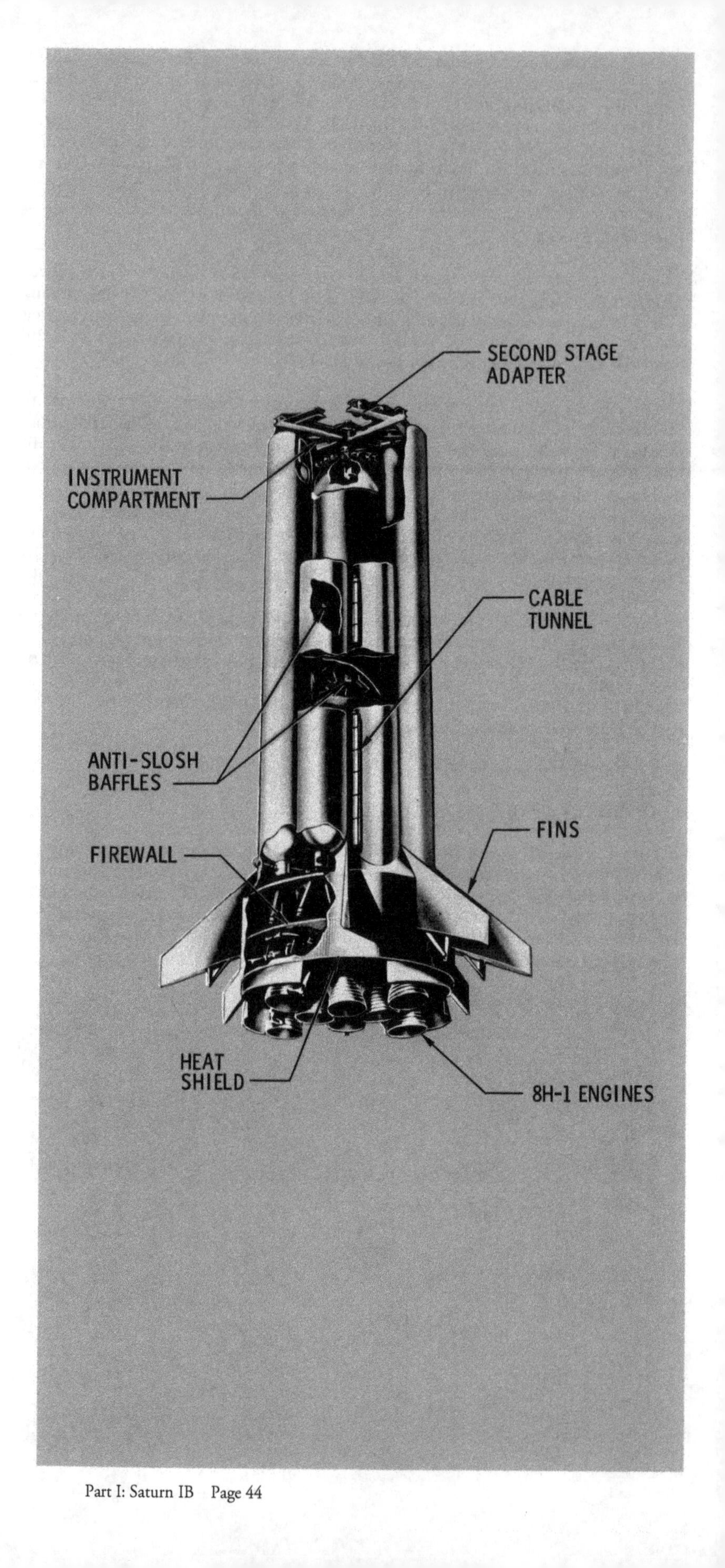

Figure III-4
S-IB STAGE INBOARD PROFILE

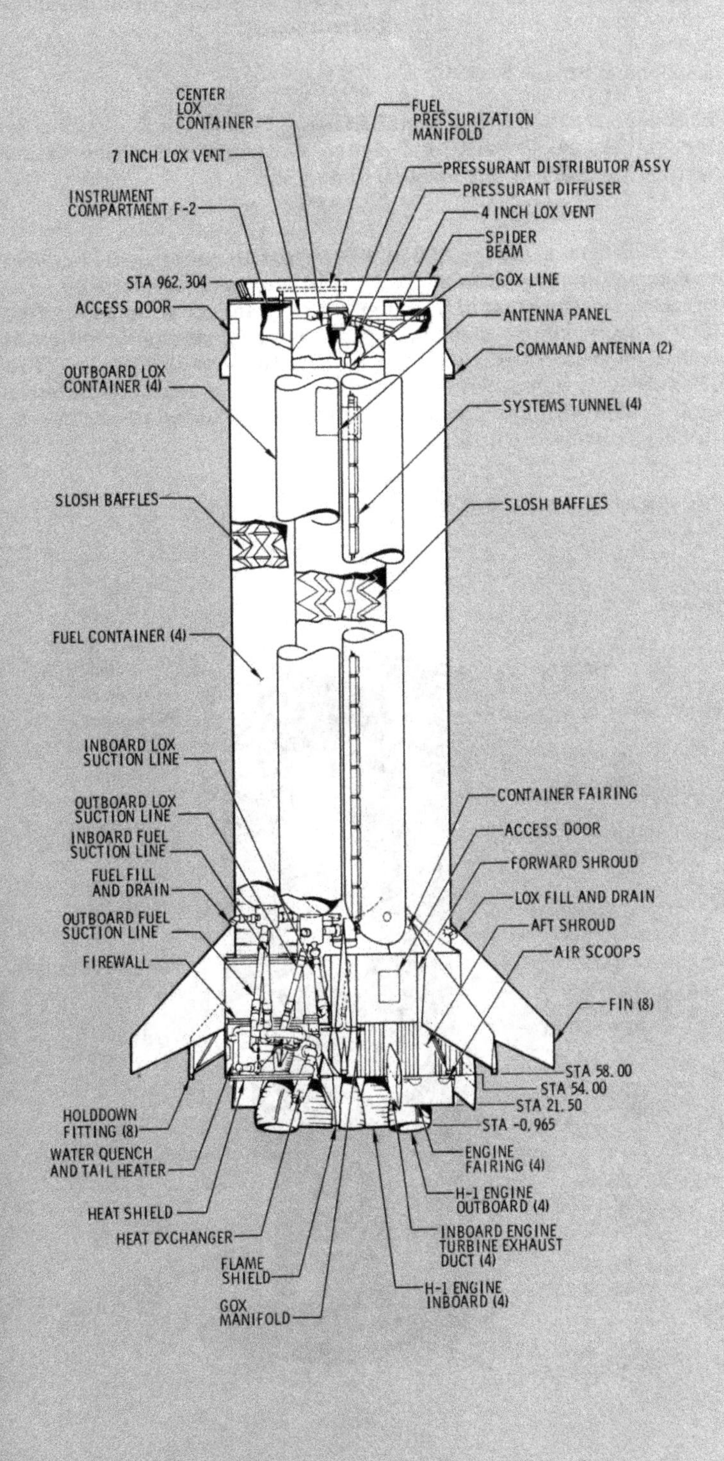

The eight Rocketdyne 200,000 lb sea level thrust H-1 engines of the S-IB stage burn approximately 150 sec, use liquid oxygen and RP-1 propellants, and lift the vehicle to over 30 nautical miles before burnout occurs. The outer ring of engines are hydraulically gimballed to provide thrust vector control in response to steering commands from the guidance system located in the Instrument Unit.

2. Second Stage (S-IVB)

The second stage of the Saturn IB is the S-IVB which is being developed by the Douglas Missile and Space Systems Division at Huntington Beach, California (Figures III-5 and -6).

The S-IVB has a single 200,000 lb thrust Rocketdyne J-2 engine that burns liquid oxygen (LOX) and liquid hydrogen (LH_2). Minor modifications and recalibration of the engine can increase thrust to 205,000 lb This engine which is planned to be incorporated on vehicle SA-205 or 206 is used in the performance calculations of Section IV. The Saturn IB/S-IVB as presently designed has a 4-1/2 hour orbital coast capability. The tankage contains 230, 000 lb of usable propellant at a LOX to LH_2 mixture ratio of 5 to 1.

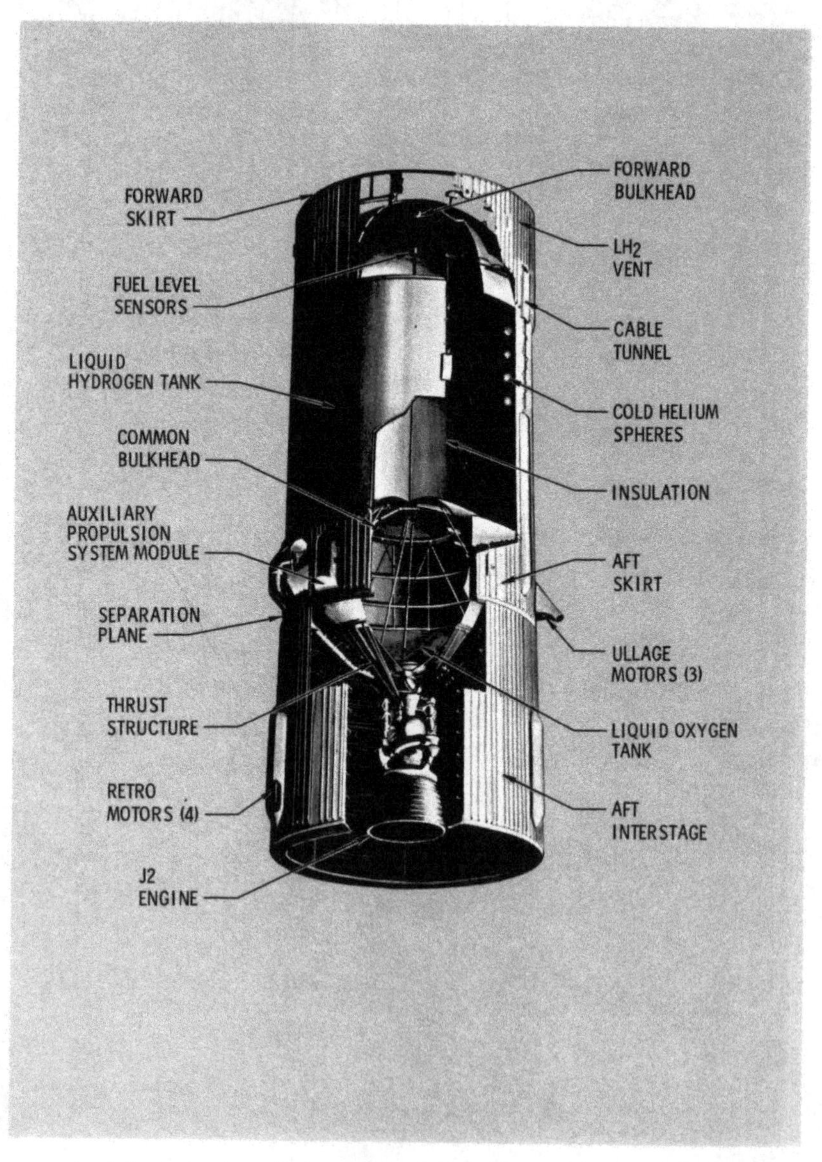

Figure III-5
SATURN IB/S-IVB STAGE

Figure III-6

SATURN IB/S-IVB STAGE
INBOARD PROFILE

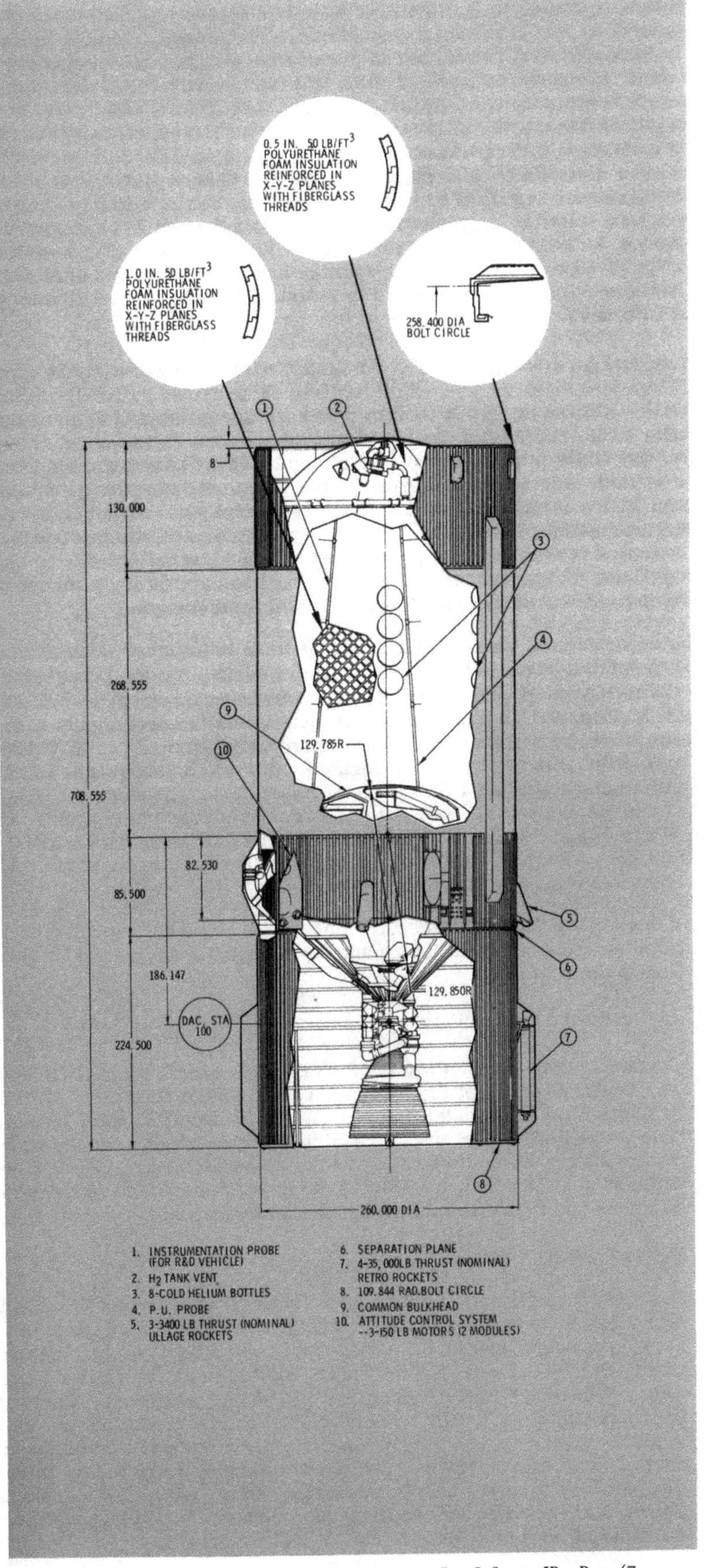

The thrust is transmitted to the stage through a skin and stringer structure shaped in the form of a truncated cone that attaches tangentially to the aft liquid oxygen dome. The hydrogen tank is internally insulated with reinforced polyurethane foam and contains a series of high pressure spheres storing gaseous helium for liquid oxygen tank pressurization. Adapter structures, referred to as the forward and aft skirt and the aft interstage, provide the necessary interfaces for mating with the payload and the lower stages. The tank structure features a waffle-like pattern on the hydrogen tank sidewall to act as a semi-monocoque load bearing member. A double walled composite structure with an insulating fiberglass honeycomb core forms the common bulkhead which separates the hydrogen and oxygen tanks. The propellant tanks have spherical end domes. Skirt and interstage structures are composed of conventional skin, external stringers and internal frames.

Pitch and yaw attitude are controlled during powered flight by gimballing the main engine. Roll control is provided by 150-lb thrust engines located in the Auxiliary Propulsion System (APS) modules. Three axis (roll, pitch, and yaw) attitude control during coast or unpowered flight is provided entirely by the APS. The signals for vehicle attitude control originate in the guidance and control system located in the instrument unit. The APS modules are located on the aft skirt assembly of the S-IVB, 180° - apart from each other and utilize nitrogen tetroxide (N_2O_4) and monomethylhydrazine (MMH) as the propellant. Each Saturn IB/S-IVB module has two 150-lb thrust roll/yaw engines and one 150-lb thrust pitch control engine.

The separation of the S-IVB from the S-IB is initiated by an explosive charge which parts the aft skirt from the aft interstage. Three 3,400-lb thrust solid propellant ullage rockets mounted on the S-IVB are then ignited and burn for 4 seconds to settle the propellants at the pump inlets by maintaining a positive acceleration. Four 35,000 lb-thrust solid retro rockets located on the aft interstage are fired simultaneously for 1.5 sec to decelerate the first stage. The J-2 engine is ignited 1.6 sec after separation signal and is at full thrust within 5 sec of the ignition signal. The J-2 burns between 480 and 500 sec. Engine shutdown is triggered by the guidance system when orbit insertion velocity is achieved.

Vehicle sequencing devices, engine control system electronics, and telemetry signal conditioning units are mounted inside the skirt section of the stage.

S-IVB Restart Capability

A second configuration of the S-IVB with a restart capability is used as the third stage of the Saturn V vehicle. The Saturn V/S-IVB configuration could provide a restart capability on the Saturn IB if desired. The Saturn IB/S-IVB would require more extensive modification to add restart than would be required to adapt the Saturn V/S-IVB to the Saturn IB. The Saturn V/S-IVB modifications include changes to the aft skirt and interstage and the electrical sequencing.

3. Instrument Unit (I.U.)

The instrument unit fabricated by International Business Machines Corporation is a 260-in. diameter by 36-in. high cylindrical section located forward of the S-IVB on both the two- and three-stage Saturn IB configurations (Figure III-7). This 3990 lb unit, which is the "nerve center" of the vehicle, contains the guidance system, the control systems and the flight instrumentation systems for the launch vehicle. Access to the inside of the S-IVB forward skirt area is provided through an I.U. door. Electrical switch selectors provide the communications link between the I.U. computer and each stage. The computer controls the mode and sequence of functions in all stages. The I.U. consists of six major subsystems:

(1) The structural system or the aluminum cylindrical body of the unit which carries the payload and supports the instrumentation.

(2) The environmental control system provides electronic equipment cooling during ground operations and throughout flight. The coolant is a 60%-40% methanol-water mixture which circulates through a series of cold plates. In flight, the absorbed heat is removed through a heat exchanger that vents boiled-off water to space. For ground operation, the system rejects heat to a thermo-conditioning servicer.

(3) The guidance and control systems provide guidance and control sensing, guidance steering computations, and control system signal shaping and summing. The shaped control signals are fed to the appropriate actuating devices on the S-IB and the S-IVB stages.

(4) The measuring and telemetry system transmits signals from the vehicle or experiment transducers during ground checkout and flight to ground command stations by various frequency bands and modulating techniques.

(5) Telemetry systems maintain contact between the vehicle and ground stations for tracking and command purposes. They consist of Azusa and C-band transponders, airborne range and orbit determination transmitters and receivers, a minitrack beacon, a radar altimeter, and a command receiver.

(6) A separate electrical system generates and distributes 28 vdc power required for operation of all of the above systems. Some of this power may be available to experimenters.

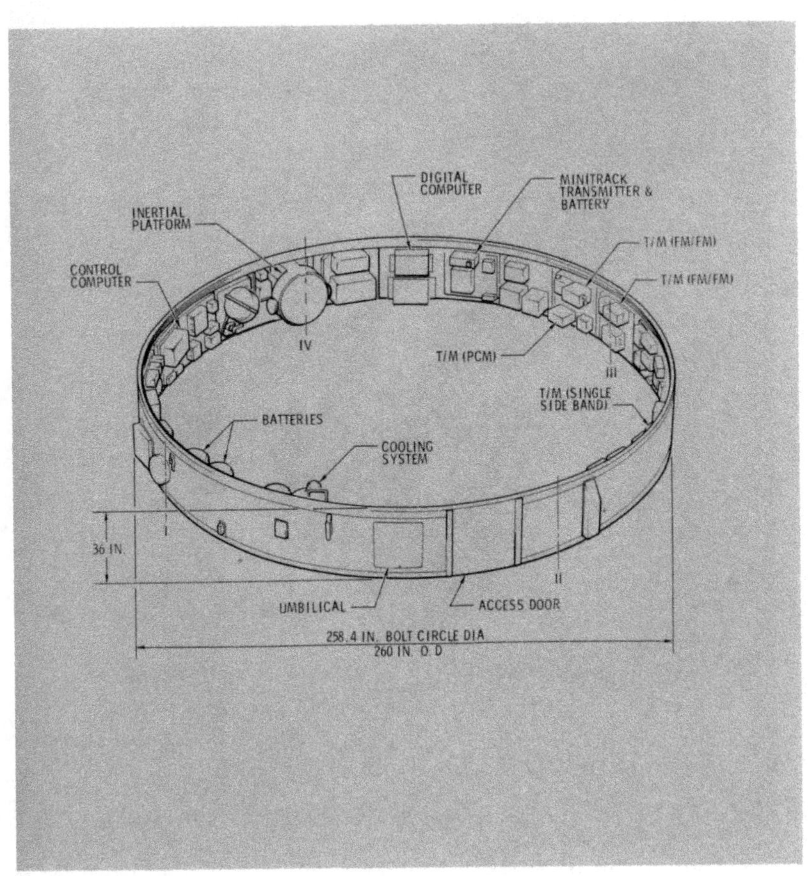

Figure III-7
INSTRUMENT UNIT

B. Three-Stage Saturn IB/Centaur

The three-stage Saturn IB will use the Centaur as the upper stage. The Centaur is manufactured by Convair Division of General Dynamics Corporation (Figures III-8 and -9). The basic stage and its later configurations, when used with Saturn IB, will be equipped with a jettisonable shroud to match the 260-inch diameter of the boost vehicle. The Centaur's main propulsion system consists of two 15,000 lb thrust Pratt and Whitney RL-10 engines which burn LOX and LH_2. The Centaur can be restarted in flight and has an extended orbital coast capability. The LH_2 tank has external insulation. With the Centaur third stage, 12,300 lb can be launched to the vicinity of the moon or up to 9,600 lb on interplanetary missions.

The Centaur has four 50-lb thrust hydrogen peroxide (H_2O_2) ullage engines used to settle propellants during S-IVB/Centaur separation, engine ignition, and again at first burn cutoff. Four 1.5-lb thrust and two 3.0-lb thrust H_2O_2 engines are used for attitude and ullage control in the orbiting vehicle.

The Centaur stage is separated from the I.U./S-IVB by four 3,400-lb thrust solid propellant retro rockets. The motors are attached to the I.U./Centaur interstage. Retro rockets for payload separation can be provided on the Centaur if required. The Centaur has its own guidance system which it uses after separation.

C. Growth Configuration of the Saturn 1B

The two-stage configuration described in this document and its performance are based on a standard operational Saturn IB. NASA and

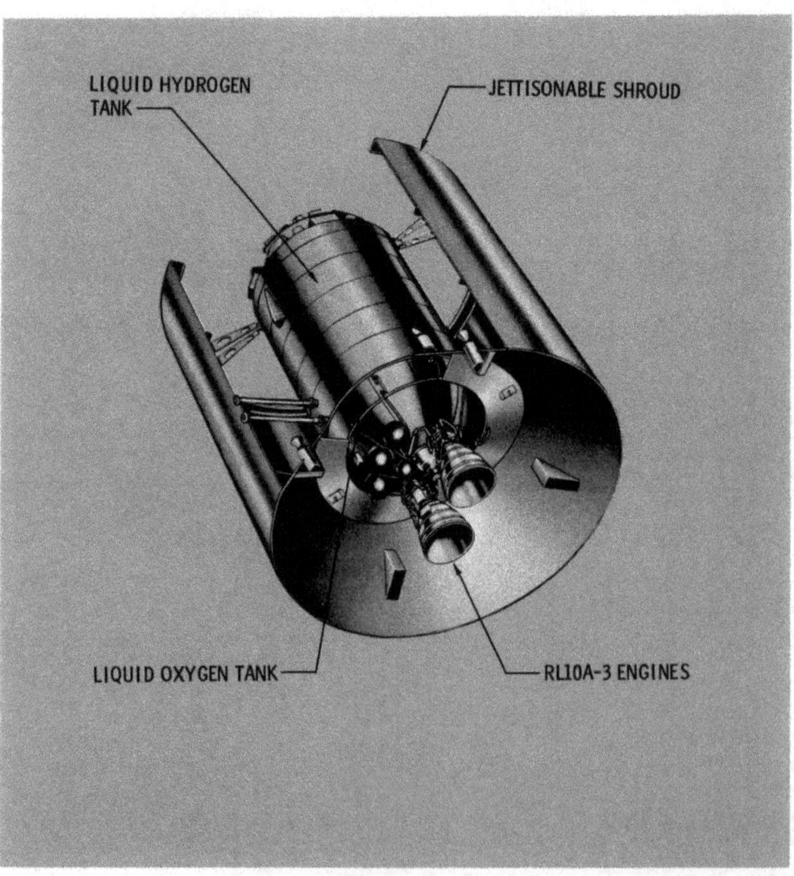

Figure III-8
SATURN IB/CENTAUR
(3RD STAGE)

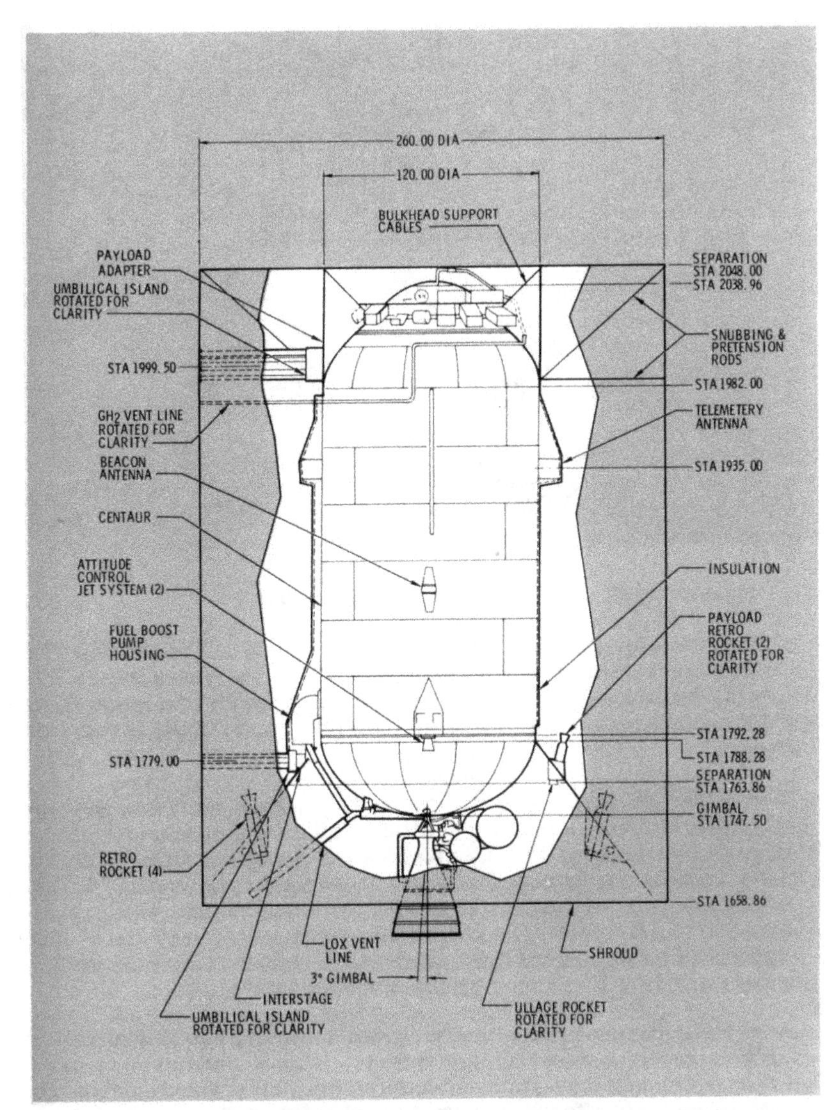

Figure III-9

CENTAUR STAGE INBOARD PROFILE WITH 260-INCH DIAMETER SHROUD

several stage contractors are continually studying means of improving the basic vehicle to achieve higher performance. One of these is presented in Figure III-10. It indicates a considerable growth potential for the existing Saturn IB which can launch over 35,000-lb payload to a 100 n.mi. orbit.

The growth configuration in Figure III-10 shows four Minuteman solid motors mounted on the Standard IB. Each solid motor provides a thrust of 200,000 lbs. The addition of these solids will provide a payload increase of 6,550 to 9,000 lbs, depending upon the ignition sequence. If one opposing pair of motors are ignited on the pad and separated 85 sec. after liftoff and the remaining motors ignited at 70 sec. after liftoff and retained throughout the flight, 6,550 lbs increase is available. If one pair of motors is ignited at liftoff and the two opposing motors ignited 10 sec. later and both jettisoned at 85 sec. after liftoff, a payload increase of 9,000 lbs can be obtained. Based on a vehicle using the first ignition sequence the payload capability of the Saturn IB/Minuteman solids is over 41,500 lbs to 100 n.mi. orbit. The same uprating technique can be applied to the Saturn IB/Centaur with corresponding increases in payload capability.

Other ways to improve the payload delivery capabilities and the versatility of the Saturn IB are under study. Payloads of 70-73,000 lbs in a 100 n.mi. orbit, approximately twice the present capability, appear to be possible.

Figure III-10

TYPICAL EXAMPLES OF GROWTH CAPABILITIES OF SATURN IB LAUNCH VEHICLE

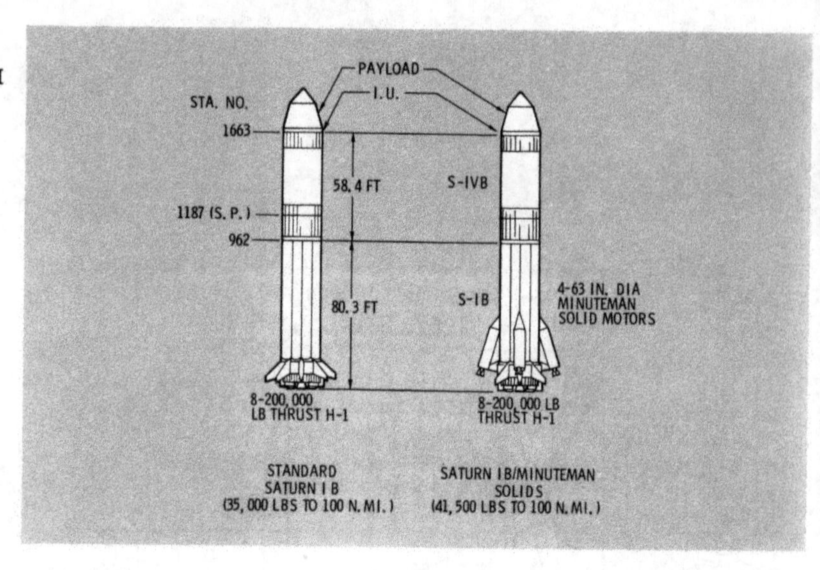

D. Man-Rating, Reliability and Quality Control

One prime objective in the design of the two-stage Saturn IB has been to provide a vehicle that can carry manned spacecraft safely. This objective is attained by providing an Emergency Detection System (EDS) and by imposing stringent Reliability and Quality Control procedures.

The EDS provides automatic warning and aborts the mission automatically if there is insufficient time for the crew to react. When sufficient time exists, the EDS provides the crew with displays enabling them to decide whether to abort or to attempt the mission. The EDS is designed to minimize the possibility of automatically aborting because of a false signal. The EDS and abort procedures are closely integrated with range safety procedures to insure that the crew can escape safely if the launch vehicle must be destroyed.

A vehicle developed for manned flights requires high reliability of each component, subsystem, and system. This is obtained by careful system design and analysis to identify all possible significant failures, categorize their effects, and point the way to their elimination through redesign and quality control. Stringent quality control standards in manufacture, fabrication, and testing insure that reliability will not be degraded by human error or by inexperience with novel manufacturing techniques. NASA documents of the NPC 200 series, (Quality Program Provisions) and NPC 250-1, (Reliability Provisions for Space System Contractors) contain the reliability requirements and quality control standards which guide payload planners.

Thorough component, subsystem and system tests are conducted in the laboratory, at the Static Test Facility and during prelaunch checkout. Post-flight data evaluation of vehicle systems serves as a tool to assess reliability for future missions. Carefully planned procedures and controls used in these tests and data correlations establish a measure of reliability and determine the level of confidence in the measure. All these factors help meet the Saturn IB reliability goal of 0.90. The reliability goal of the S-IVB stage is 0.95 at a 90% confidence level. Experiment payloads being carried on the Saturn IB will be given the same attention to insure the same high probability of success.

Unmanned payloads will benefit from the stringent quality control procedures and the high reliability required of a man-rated launch vehicle. Auxiliary and unmanned prime payloads must meet these same requirements.

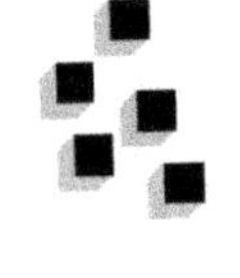

A. Two-Stage Mission Profile

On two-stage missions the launch sequence starts with the ignition of all eight first-stage engines. The vehicle rises vertically for approximately 15 sec to clear the umbilical tower and rotates so that the pitch plane is aligned with the desired launch azimuth. A sequence of pitch maneuvers then starts the vehicle down range. The pre-programmed pitch maneuvers are designed to follow a ballistic trajectory (zero angle of attack). At 140.3 sec after lift-off, the center four engines are shut down, and six seconds later the other four are cut off. Five and one half seconds are then required to separate the empty S-IB stage and reach 90% thrust on the S-IVB stage.

The second stage is controlled by an iterative guidance scheme based on the calculus of variations which minimizes the propellant burned in reaching the desired burnout velocity and location. The launch phase of two-stage missions ends when the S-IVB has injected itself and its payload into the desired orbit approximately 620 sec after lift-off.

B. Three-Stage Mission Profile

For three-stage missions, the shroud enclosing the Centaur is jettisoned at an altitude of approximately 350,000 feet altitude 192 sec after lift-off. The S-IVB burns out at approximately 610 sec after which separation and ignition of the Centaur stage occurs. Centaur at this point may utilize one of two techniques to complete earth orbit missions; direct ascent or Hohmann transfer. In the first case, one burning phase is required. In the second, two or more burning phases are required depending on whether or not an elliptical transfer orbit or a circular low earth parking orbit is to be established at the first burnout. If the former, one additional burn phase will be required to establish the desired orbit. If the latter, two or more burn phases may be needed to establish the desired orbit shape and orientation. Earth escape missions may be accomplished by either direct ascent (one Centaur burn phase) or from a low earth parking orbit (two Centaur burns).

Table IV-I gives the weight breakdown of the Saturn IB stages for both two- and three-stage launch vehicles. The S-IVB and Centaur propellant figures shown are tank capacities. The actual amount burned would depend on the specific mission. In determining performance of the three-stage vehicle, a constant S-IVB propellant loading of 220,000 lb was used. This is not necessarily optimum for all missions and some gain in performance may be achieved by optimizing this parameter for a specific mission.

Figures IV-1 and IV-2 present the circular orbit payload capabilities of two- and three-stage vehicles for direct ascent missions to various orbit altitudes and inclinations when launched from the ETR. Payload capability is shown in Figure IV-3, for the case of a due east launch. These curves are based on the assumption that the launch site is in the plane of the desired orbit and no trajectory plane-changing "dog leg" maneuvers are used. The approximate sector of probable launch azimuths at ETR without "dog legging" are between 40 and 140 degrees. To achieve orbit inclinations greater than approximately 55 degrees, (launch azimuths less than 40 degrees or higher than 140 degrees) would require range safety approval and a plane-change with a resultant decrease in payload. To launch a two-stage vehicle on a 140° azimuth and dog leg to 100 n.mi. polar orbit would reduce payload to 23,800 lb from 27,900 lb for the coplanar case (see Figure IV-1).

Figures IV-4 through IV-7 show the payload capability of the two- and three-stage vehicles to various elliptical orbits for due east launches out of ETR and polar orbits out of the Western Test Range (WTR).

The interplanetary capability of the three-stage vehicle is presented in Figure IV-8. These data are based on a due east launch out of ETR to a 100 n.mi. parking orbit and require two Centaur burns.

Some representative trajectory time histories are shown in Figures IV-9 through IV-15 for the two- and three-stage vehicle launched due east from ETR to a 100 n.mi. circular orbit.

Table IV-1
Vehicle Weight Summaries
(Projected Operational Weights)

	2 Stage Weights, Lb	3 Stage Weights, Lb
Dry S-IB	85,500	85,500
Residuals	11,810	11,810
S-IB/S-IVB Interstage	6,550	6,550(2)
Separation/Start Losses	1,180	1,180
S-IB Propellant	882,430	882,430
S-IB Expendables (Frost, Oils, Misc.)	3,660	3,660
Dry S-IVB	23,120	23,120(2)
Residuals	1,920	1,920
S-IVB Propellant	228,700	220,000
Flight Performance Reserve	1,300	(1)
Instrumentation Unit	3,990	3,990
Centaur Shroud	-	5,600
S-IVB/Centaur Interstage	-	500
Centaur Start Losses	-	40
Centaur Orbit Losses	-	120
Centaur Propellant	-	30,000
Centaur Dry	-	5,350(3)
Centaur Residuals	-	440
	1,250,160 lb.(4)	1,282,210 lb.(4)

Note: (1) Flight performance reserve for the three-stage configuration is in the Centaur and is calculated as 3/4% of the total velocity requirement converted impulsively to propellant. Programmed mixture ratio effects not included.

(2) These values based on two stage weights. Due to higher loads of three-stage booster the S-IVB may require an additional 360-lb in S-IVB structure and S-IB/S-IVB interstage. Dry weight includes jettisonable items.

(3) Centaur dry weight included its own guidance for use after Centaur separation. Improved insulation would result in a 1300 lb reduction in dry weight, i.e. a 1300 lb payload increase.

(4) Weights do not include a launch escape tower.

Figure IV-1
SATURN IB CIRCULAR
ORBIT CAPABILITY

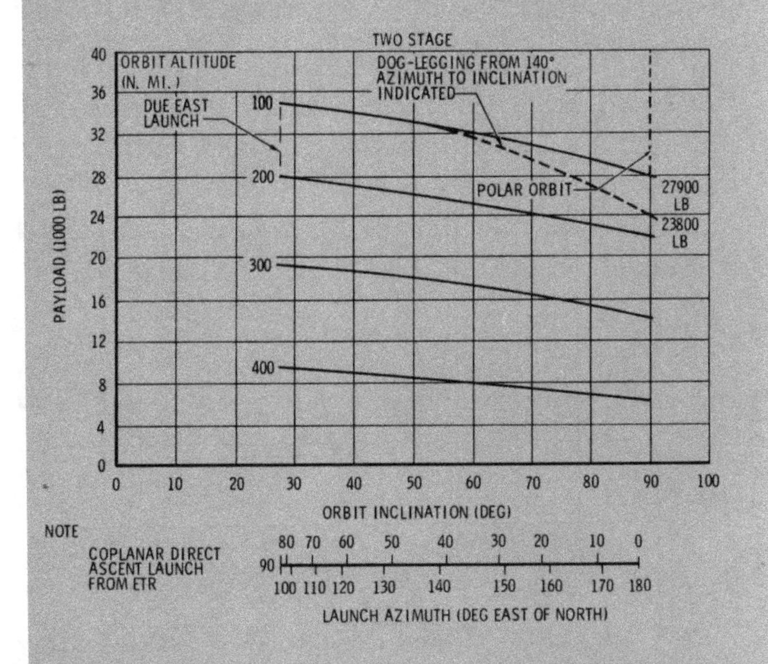

Figure IV-2

SATURN IB/CENTAUR CIRCULAR ORBIT CAPABILITY

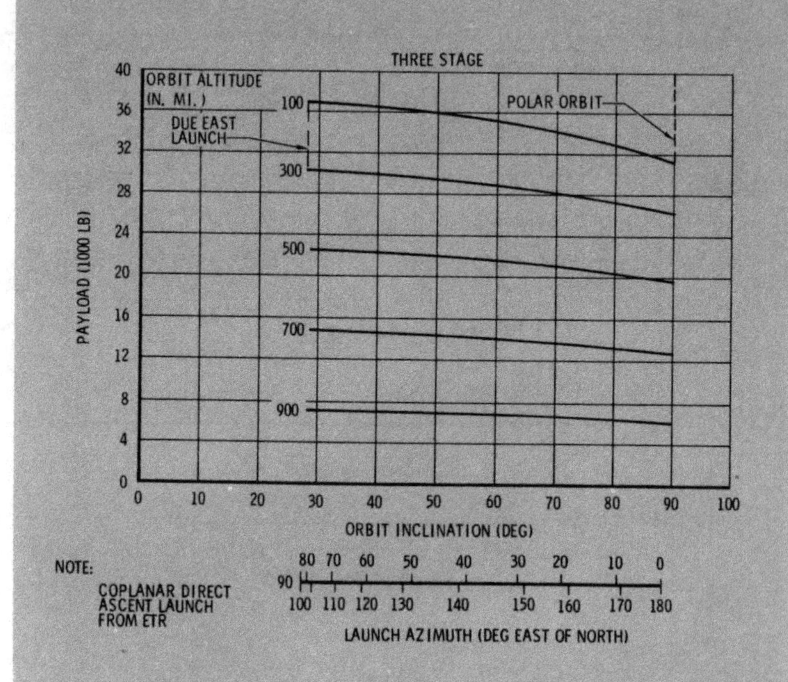

Figure IV-3
SATURN IB/CENTAUR HOHMANN TRANSFER PAYLOAD CAPABILITY FROM 100 N.MI. ORBIT

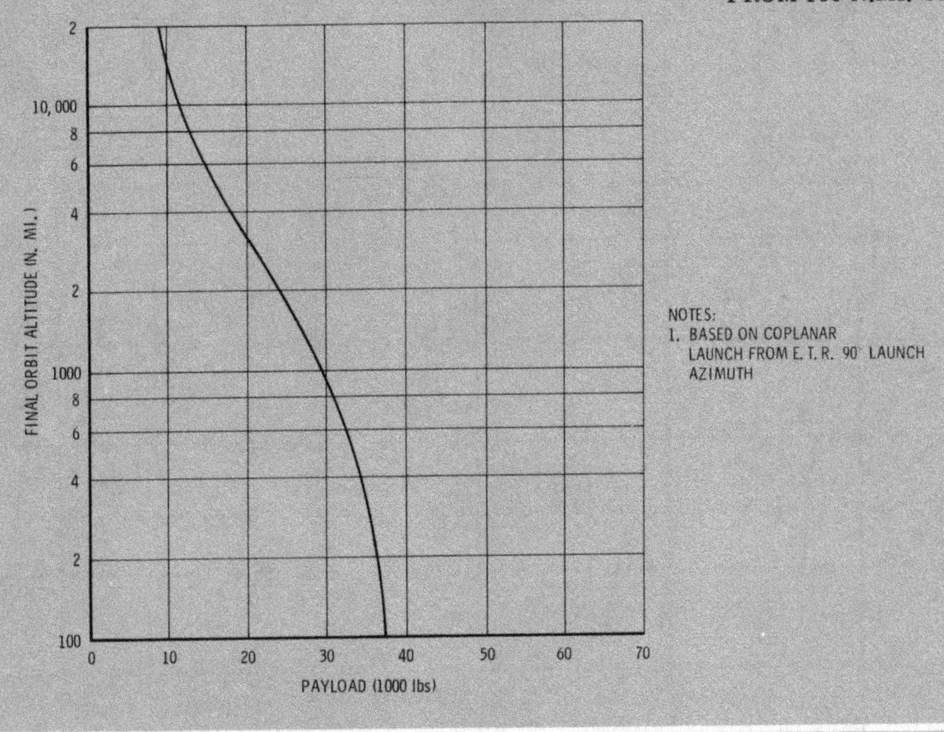

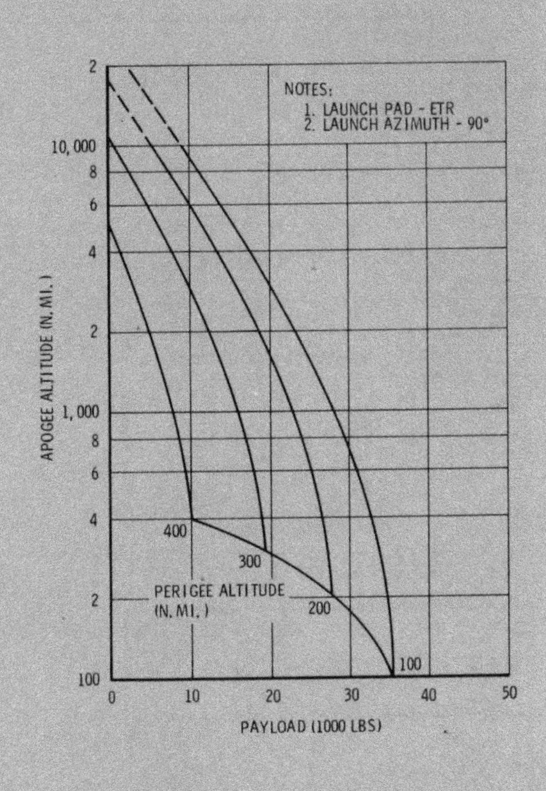

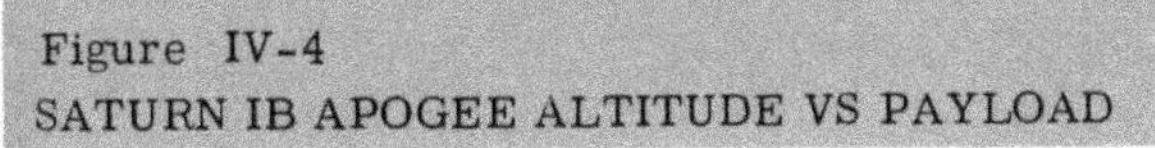
Figure IV-4
SATURN IB APOGEE ALTITUDE VS PAYLOAD

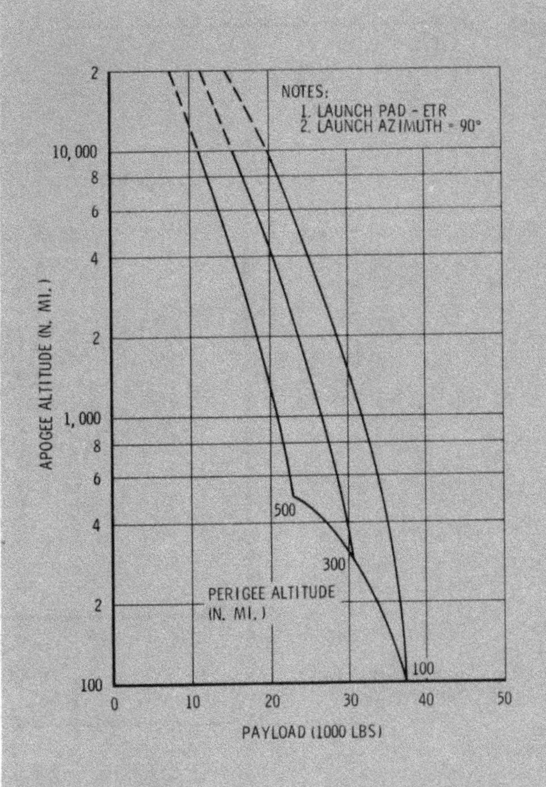

Figure IV-5
SATURN IB/CENTAUR APOGEE ALTITUDE VS PAYLOAD

Figure IV-6
SATURN IB APOGEE ALTITUDE VS PAYLOAD

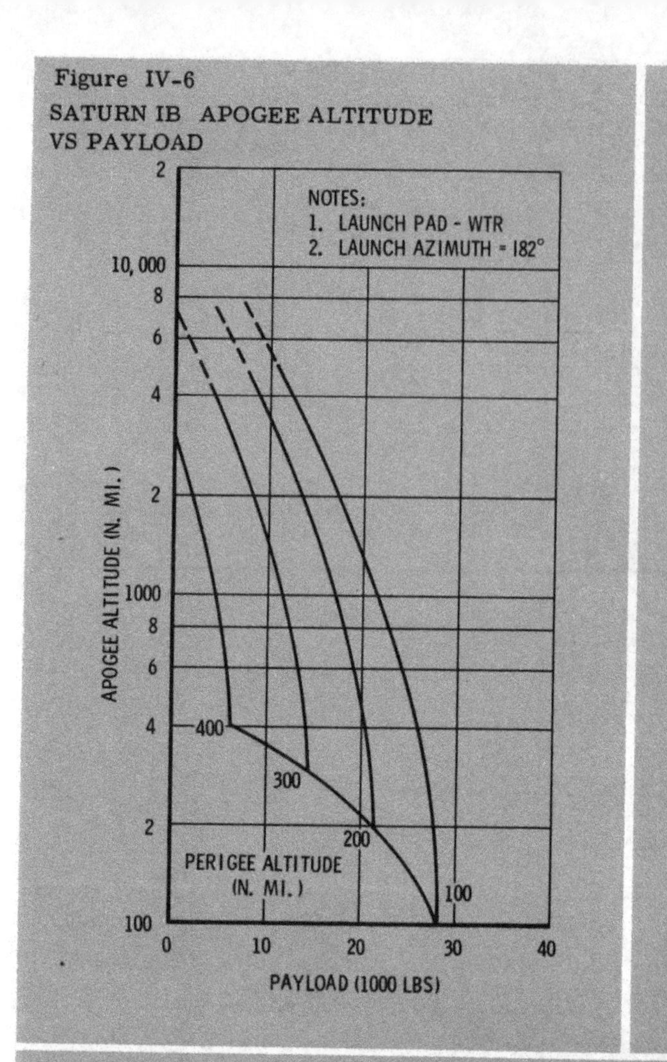

Figure IV-7
SATURN IB/CENTAUR APOGEE ALTITUDE VS PAYLOAD

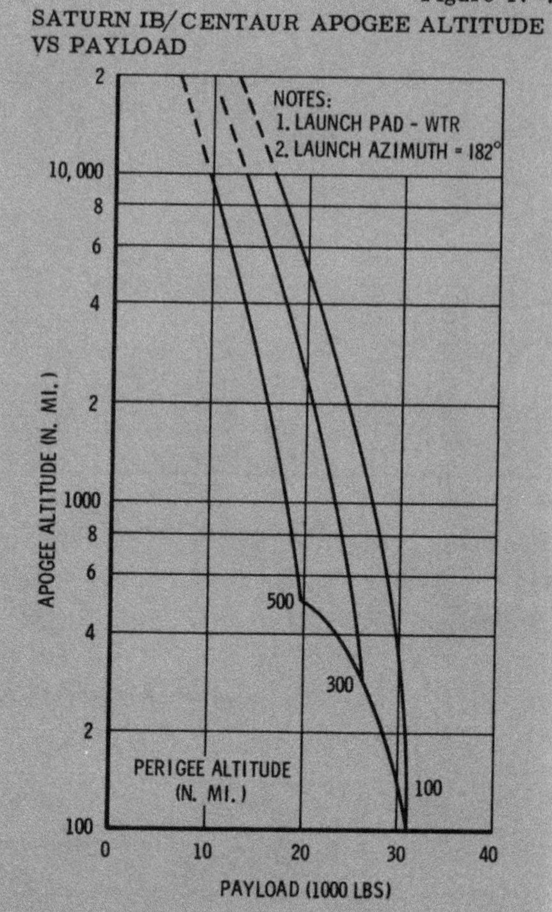

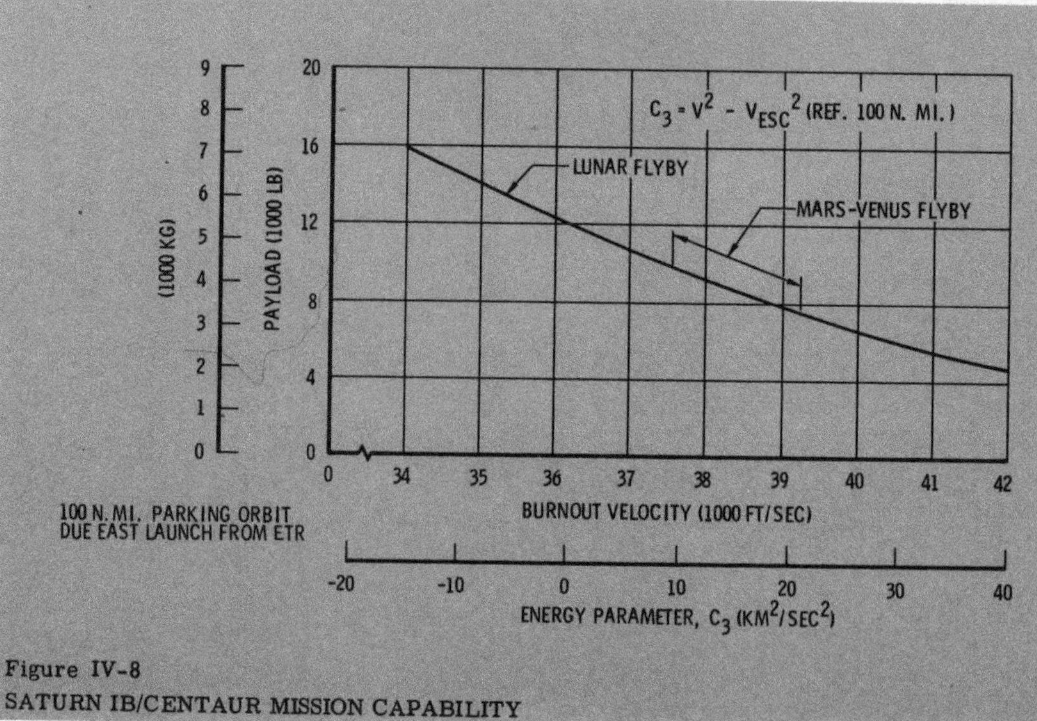

Figure IV-8
SATURN IB/CENTAUR MISSION CAPABILITY

Figure IV-9
SATURN IB INERTIAL
VELOCITY VS FLIGHT TIME

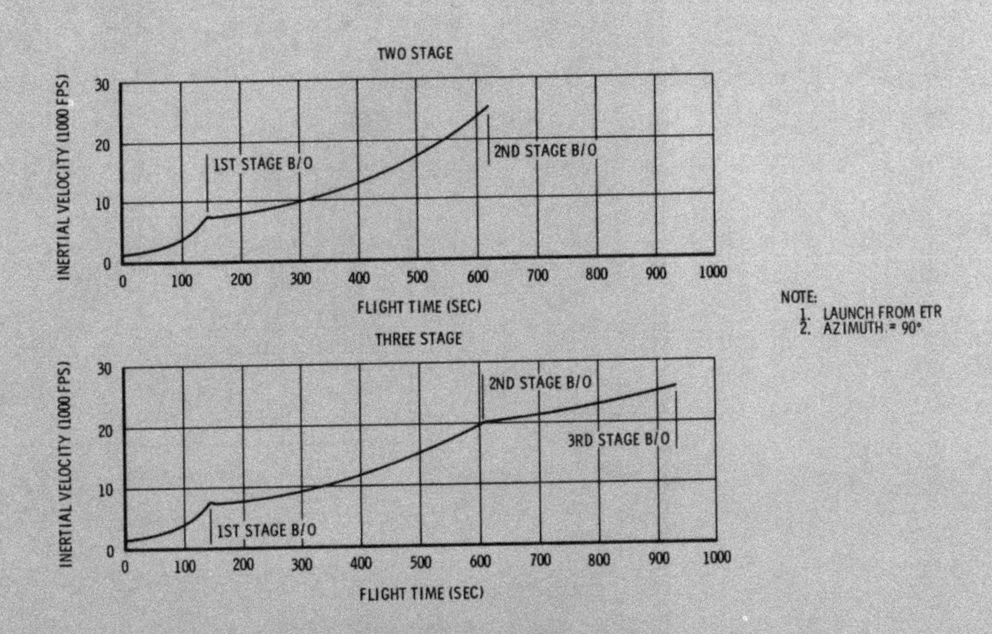

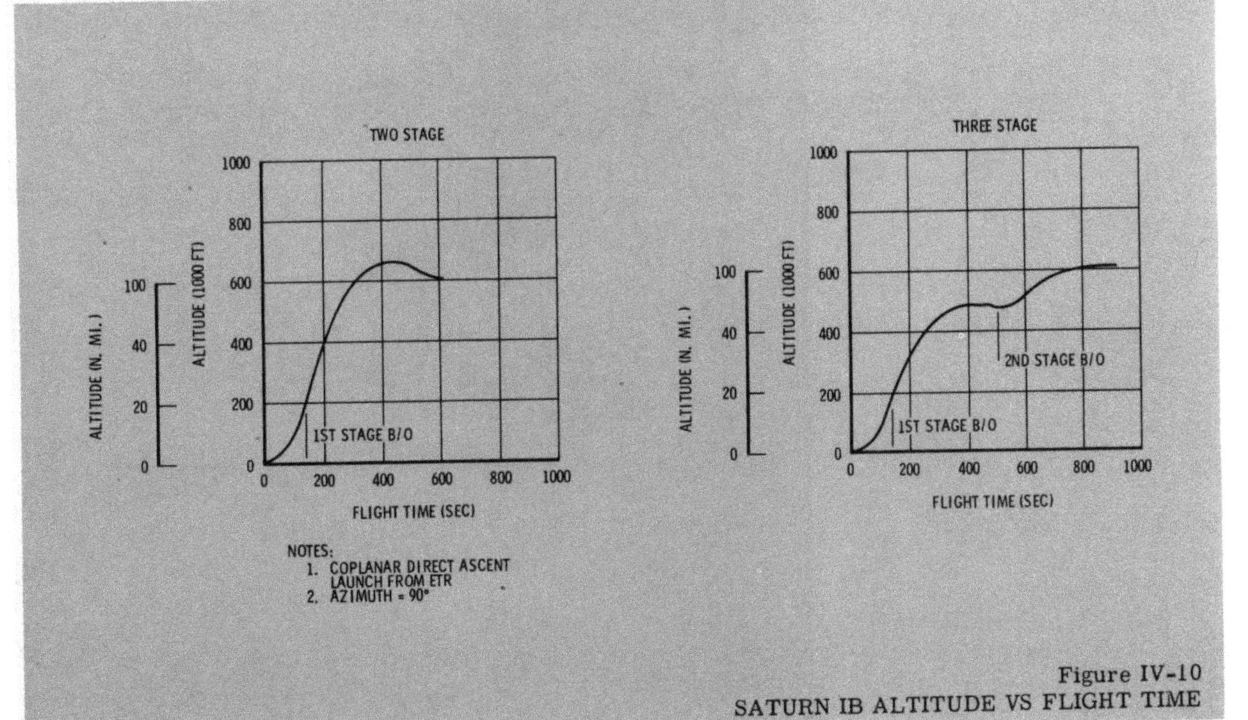

Figure IV-10
SATURN IB ALTITUDE VS FLIGHT TIME

Figure IV-11
SATURN IB INERTIAL FLIGHT PATH ANGLE VS FLIGHT TIME

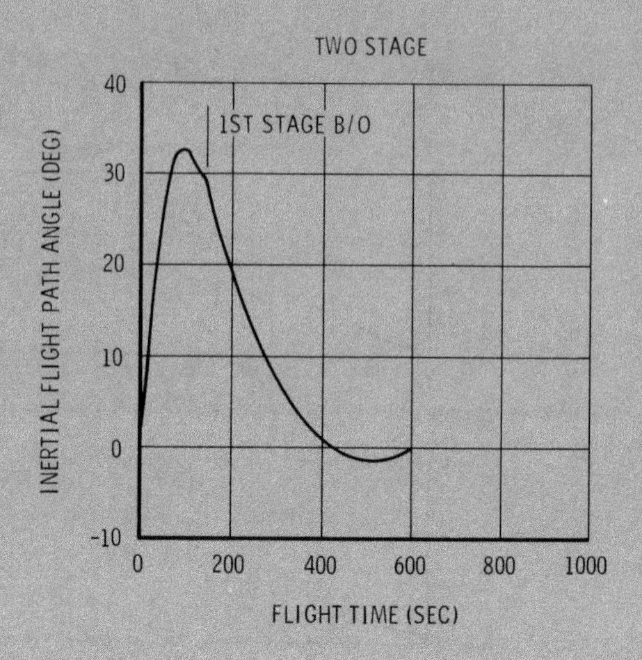

NOTES:
1. COPLANAR DIRECT ASCENT LAUNCH FROM ETR
2. AZIMUTH = 90°

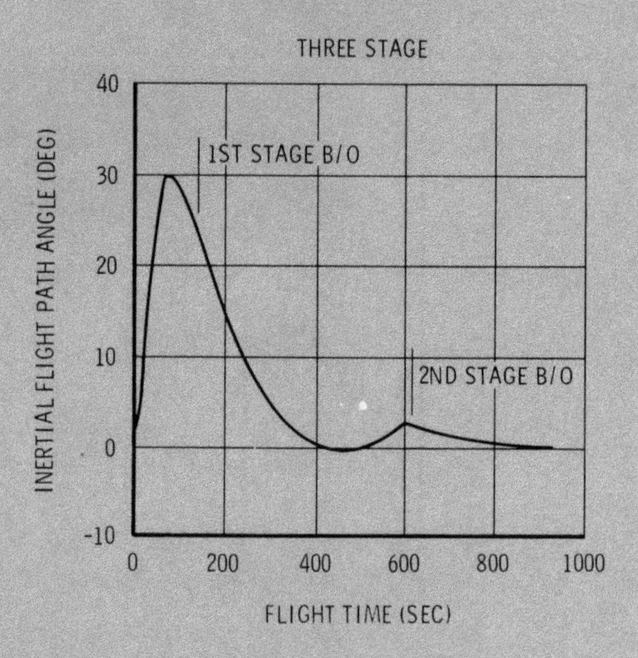

Figure IV-12
SATURN IB AXIAL
ACCELERATION VS
FLIGHT TIME

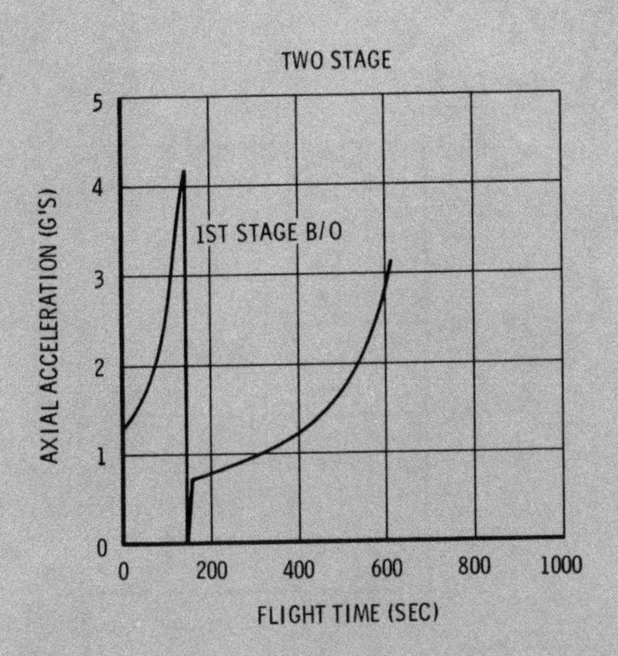

NOTES:

1. COPLANAR DIRECT ASCENT LAUNCH FROM ETR
2. AZIMUTH = 90°

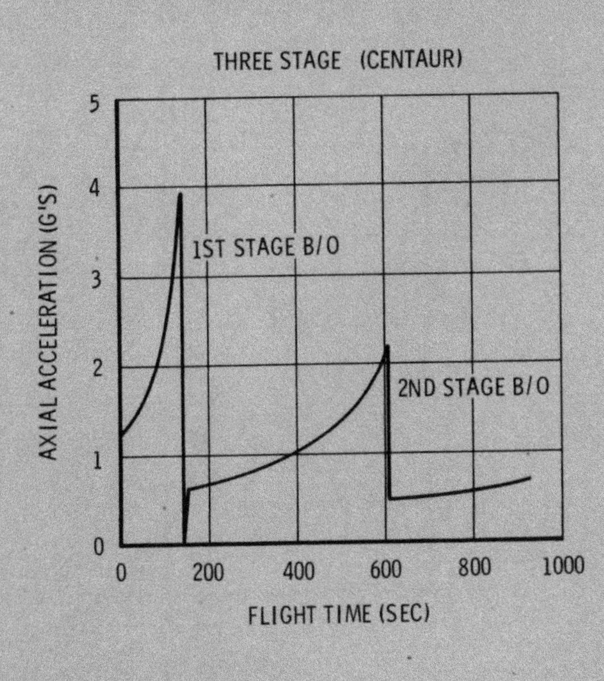

Figure IV-13
SATURN IB DYNAMIC PRESSURE
VS FLIGHT TIME

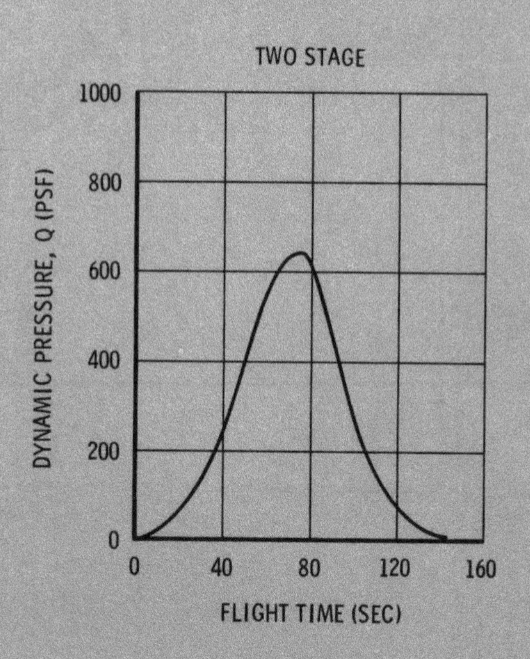

NOTES:

1. COPLANAR DIRECT ASCENT LAUNCH FROM ETR
2. AZIMUTH = 90°
3. ORBIT ALTITUDE - 100 NM

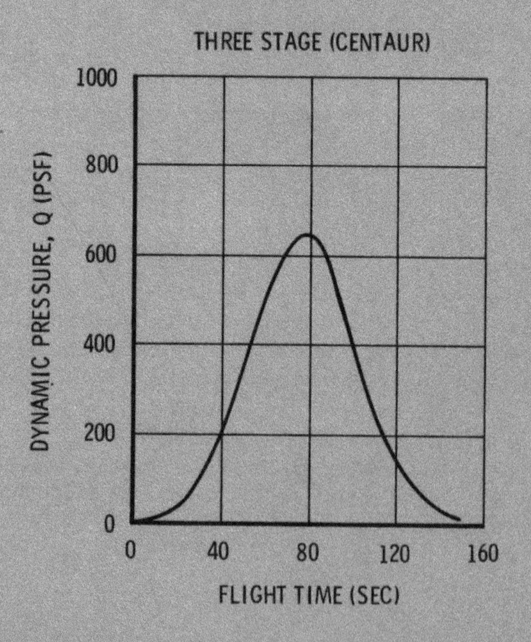

Figure IV-14
SATURN IB MACH NUMBER
VS FLIGHT TIME

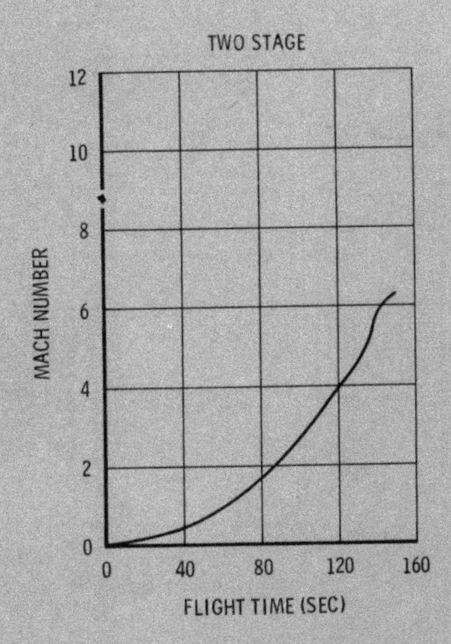

NOTES:
1. COPLANAR DIRECT ASCENT LAUNCH FROM ETR
2. AZIMUTH = 90°
3. ORBIT ALTITUDE - 100 (N. MI.)

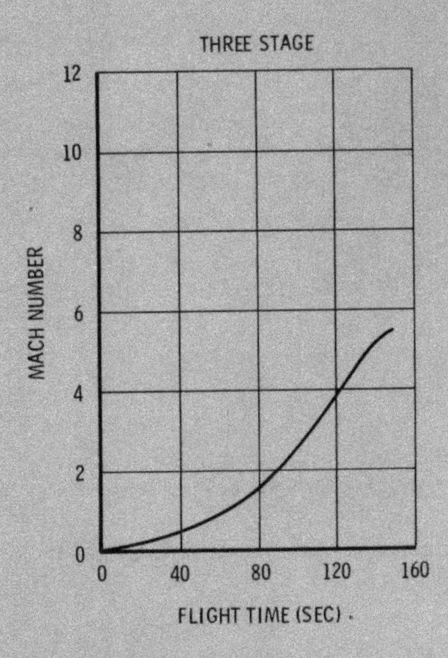

Figure IV-15
SATURN IB ALTITUDE VS RANGE

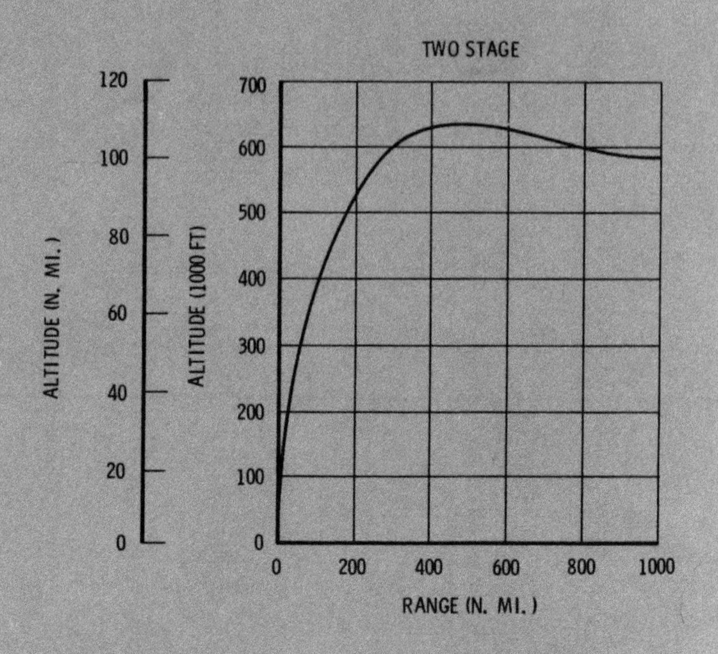

NOTES:
1. COPLANAR DIRECT ASCENT LAUNCH FROM ETR
2. AZIMUTH = 90°

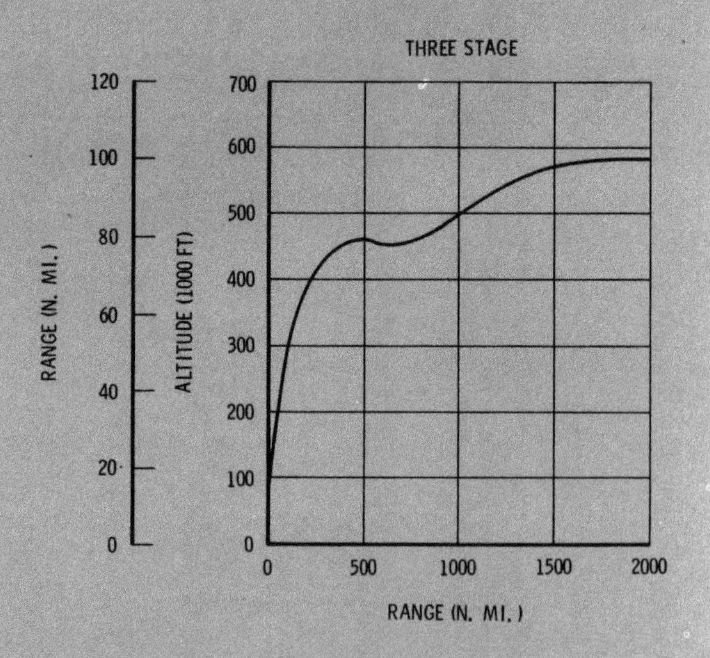

Saturn IB can carry:

Experiments in the space sciences . . . . .

Engineering tests that require the actual environment of space . . . . .

Prime missions that require a launch system providing great weight-lifting ability or high velocity.

MISSILE & SPACE SYSTEMS DIVISION
DOUGLAS AIRCRAFT COMPANY, INC.
SANTA MONICA/CALIFORNIA

DOUGLAS

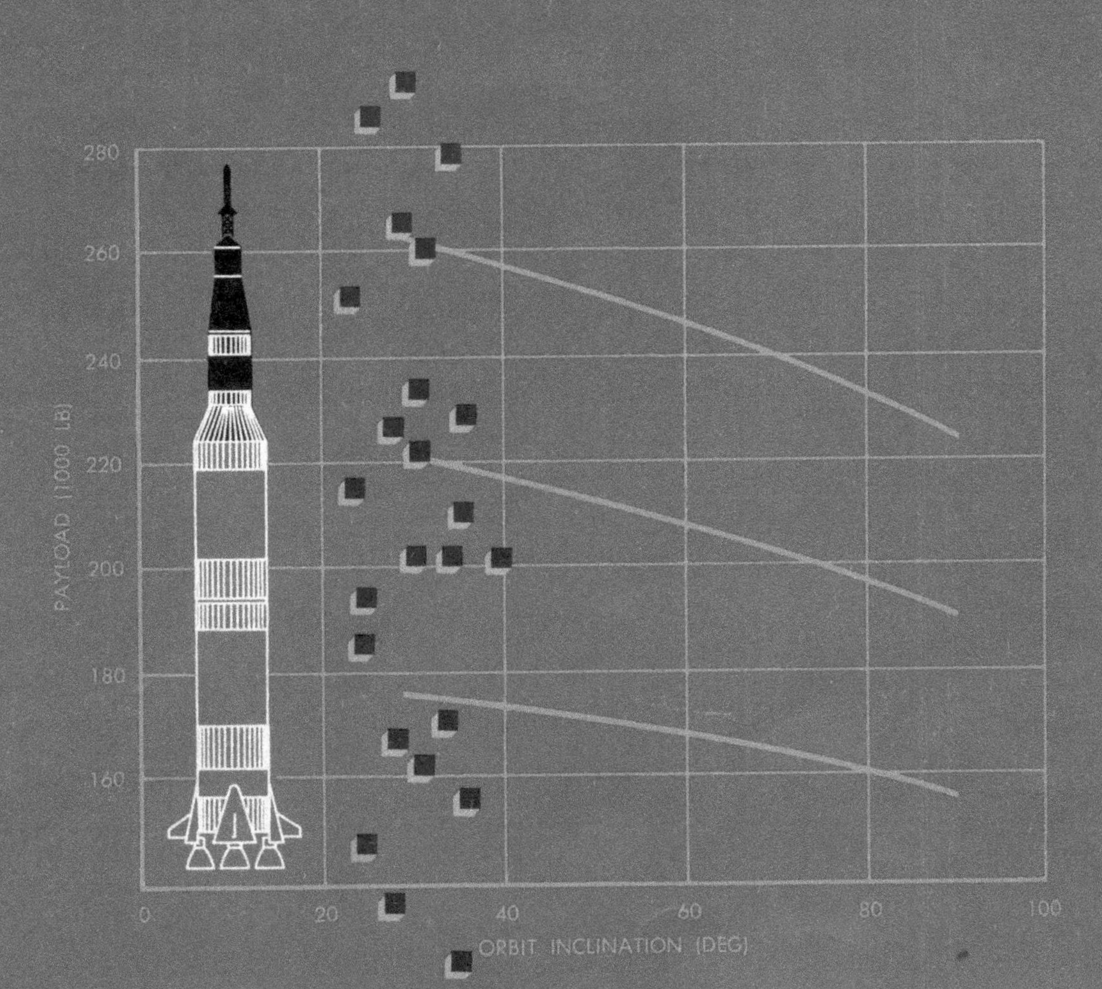

MISSILE & SPACE SYSTEMS DIVISION
DOUGLAS AIRCRAFT COMPANY, INC.
SANTA MONICA/CALIFORNIA

SATURN V

PAYLOAD PLANNER'S GUIDE

November 1965

Douglas Report SM-47274

Prepared By: L. O. Schulte

Development Planning - Saturn

Payload Applications

Approved By: F. C. Runge

Program Manager - Saturn

Payload Applications

Approved By: T. J. Gordon

Director of Advance Saturn and

Large Launch Vehicles

DOUGLAS MISSILE & SPACE SYSTEMS DIVISION

HUNTINGTON BEACH, CALIFORNIA

FOREWORD

This guide has been prepared by Douglas to acquaint payload planners with the capability of the Saturn V Launch Vehicle and to assist them in their initial payload/launch vehicle planning. This guide is not an offer of space aboard Saturn. Only NASA can commit experiments to this vehicle. This book attempts to show methods by which Saturn could accommodate payloads of various weights, volumes and missions. You will see that the capabilities of this vehicle permit a wide spectrum of assignments, including scientific, technological as well as operational type payloads.

A similar guide has been prepared on the capabilities of the Saturn IB Launch Vehicle. This book, called the Saturn IB Payload Planner's Guide, Douglas Report No. SM-47010 is available upon request.

Requests for additional information may be addressed to:

Mr. Fritz Runge, Program Manager
Saturn Payload Applications
Douglas Missile and Space Systems Division
5301 Bolsa Avenue
Huntington Beach, California 92646

Telephone 714-897-0311

SM - 47274

SATURN V PAYLOAD PLANNER'S GUIDE

TABLE OF CONTENTS

Part II: Saturn V Page 72

The Saturn V is a three stage launch vehicle under development by the NASA to support the Apollo Lunar Landing mission.

The Saturn V Vehicle will also be used to achieve many other objectives related to the national goal of lunar exploration and space flight. Certainly the development of future space stations and inter-planetary spacecraft will rely heavily upon hardware and techniques developed in the Apollo.

The Saturn V Vehicle is designed to launch very large manned and unmanned payloads into space. Each of the stages are now on the production line and progressing on schedule. The initial flight tests for the Saturn V Vehicles will be in early 1967 and will be capable of injecting over 261,000 pounds of payload into a 100 nautical mile circular earth orbit. Since the S-IVB third stage actually goes into orbit along with the payload, the total weight in orbit is nearly 300,000 pounds. In the future, a Saturn V with a high-energy fourth stage could provide an effective configuration for high velocity missions; for example, approximately 12,000 pounds could be accelerated to a hyperbolic excess velocity of 45,000 feet/second.

The Saturn class of vehicles thus constitutes a great national resource which is destined to serve the launch vehicle needs of a wide variety of future manned and unmanned space missions.

INTRODUCTION

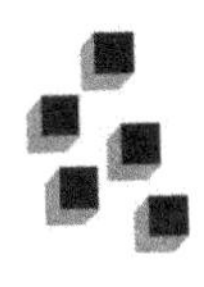

This Payload Planner's Guide is intended as a starting point for engineers, scientists, and executives who are planning to conduct engineering tests, space science experiments, or, operational missions. It outlines, for the payload planner, the technical information and procedures with which large prime, or small auxiliary payloads can be effectively integrated and flown on the vehicle. The payload planner will find here the characteristics of the Saturn V launch vehicle, its performance, the accommodations it offers to potential experimenters, suggested procedures to be followed in obtaining support for the experiment, approximate flight schedules and engineering data needed to initiate the design of a payload. To planners of prime payloads, the guide offers four protective shroud designs. To auxiliary payload planners, it presents several payload accommodation concepts for identifying and describing volumes in the Saturn V where such payloads could be installed. Environmental data and payload weight limitations for each payload volume are provided.

The Saturn V performance capabilities are included for payload flight planning. The major subsystems of the launch vehicle and their relation to the payloads are described. A concept-to-flight chronology of events is presented to support payload/launch vehicle system planning on the part of prospective users.

The Douglas Missile and Space Systems Division will be pleased to discuss the planning, support, operation, and data evaluation involved in the flight of any payload on Saturn V.

SATURN V CONFIGURATIONS

The three-stage configuration of the Saturn V, depicted in Figure 1-1, is the basis for the data presented in this guide. The three stage Saturn V with the Apollo spacecraft is about 363 feet tall and weighs nearly 3200 tons. A possible four-stage configuration is described in Section V.

FIRST STAGE (S-IC)

The S-IC stage is 138 feet tall and 33 feet in diameter. The propellants, liquid oxygen (LOX) and RP-1 (special kerosene fuel), are stored in two separate tanks with the fuel in the lower tank. Five 1.5 million pound thrust F-1 engines are used to generate a total of 7.5 million pounds of thrust at lift off, and propel the vehicle to an altitude of 30 nautical miles in 150 seconds. Four of the engines are hydraulically gimballed to provide thrust vector control in response to steering commands from the guidance system located in the Instrument Unit. The first stage is separated from the second by eight 80,000 pound thrust solid rocket motors.

SECOND STAGE (S-II)

The S-II stage is 81.5 feet tall and 33 feet in diameter. The propellants, liquid oxygen (LOX) and liquid hydrogen (LH_2), are stored in two tanks separated by a common bulkhead with the LOX in the lower tank. Five 205,000 pound thrust J-2 engines propel the vehicles to a burnout altitude of 90 to 100 nautical miles depending upon the mission. Four of the engines are gimballed for control during flight, similar to the S-IC. Eight 22,900 pound thrust solid motors are fired to ullage the propellants for engine start. The second stage is separated from the third by four solid rocket motors, each of which produces 35,000 pounds of thrust for 1.5 seconds. The interstage which mates the S-IVB to the S-II remains with the S-II.

THIRD STAGE (S-IVB)

The S-IVB stage is 58.5 feet tall and is about 22 feet in diameter. The S-IVB is powered by a single Rocketdyne J-2 engine that burns liquid oxygen and liquid hydrogen to provide a thrust of 205,000 lb at an engine mixture ratio of 5/1. During flight, the main engine is hydraulically gimballed in pitch and yaw to provide thrust vector control in response to commands from the instrument unit. Powered roll control is provided by fixed position hypergolic engines located in two auxiliary propulsion system (APS) modules mounted 180° apart on the aft skirt. Three axis (roll, pitch and yaw) attitude control during coast is also provided by the APS. Two solid-propellant ullage rockets are fired at stage separation, just prior to ignition of the J-2 engine, to assure that the propellant is settled in the bottom of the tanks during the start phase. After second stage separation, the J-2 engine on the S-IVB stage ignites and propels the payload to the desired altitude. The S-IVB as presently designed has a 4-1/2 hour orbital coast plus a 2 hour translunar coast capability. Two 72 lb-thrust hypergolic

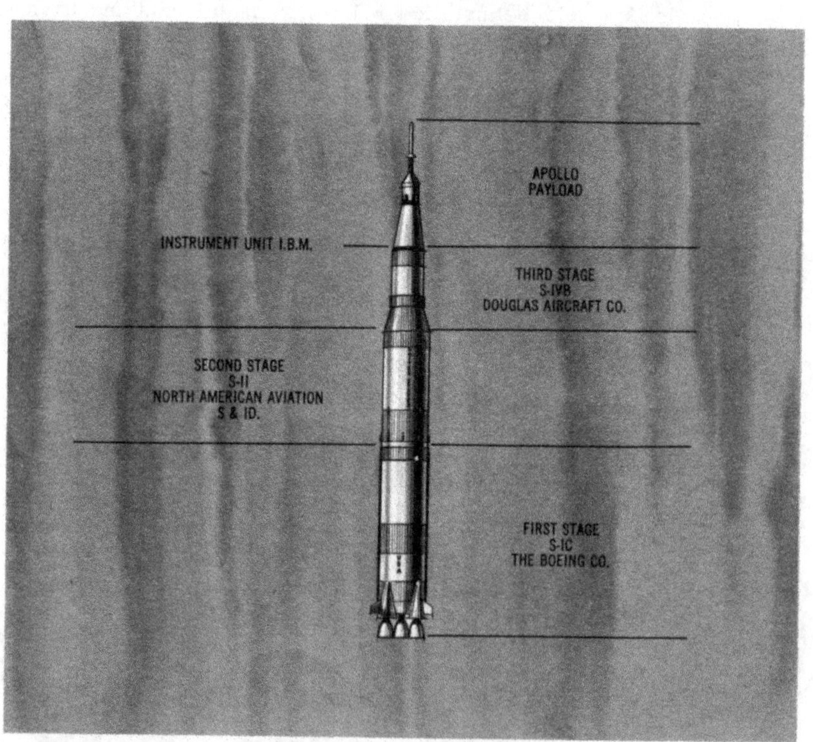

Figure I-1
SATURN V THREE
STAGE LOR
CONFIGURATION

engines in the APS modules, are fired during the first shutdown of the J-2 engine to control the position of the propellants and again, prior to the second J-2 start, to position the propellants during chilldown and restart of the main engine.

INSTRUMENT UNIT (I. U.)

The Instrument Unit houses the guidance and control systems and the flight instrumentation systems for the Saturn V launch vehicle. Specifically, the I. U. contains electrical, guidance and control, instrumentation, measuring, telemetry, radio frequency, environmental control, and emergency detection systems.

SATURN V CAPABILITY

Saturn V has the capability to perform a broad spectra of manned and unmanned space missions and can carry large prime and auxiliary payloads as summarized in Figure I-2 and presented in detail in Sections IV and V.

The major advantages of utilizing the Saturn V are:

- Largest orbital payload capability of any vehicle in the world (about 261,000 lb to 100 n. mi.).
- Large diameter payload volume.
- Large escape payload capability (about 98,000 lb).
- Large synchronous orbit payload capability (over 72,000 lb to a 20,000 n. mi. orbit with a 28.5° inclination and about 62,000 lb to an equatorial synchronous orbit).
- Low transportation costs per pound of payload in orbit.
- Flight proven systems and subsystems.
- Man-rated systems.
- Production stages.
- Complete and existing manufacturing, test and launch facilities.
- Flexibility of planning two, three or four stage configurations.
- Associated NASA data acquisition and tracking networks are operational.
- Auxiliary payload volumes, weight, power, and data channels may be available.
- Growth potential of the vehicle is considerable.

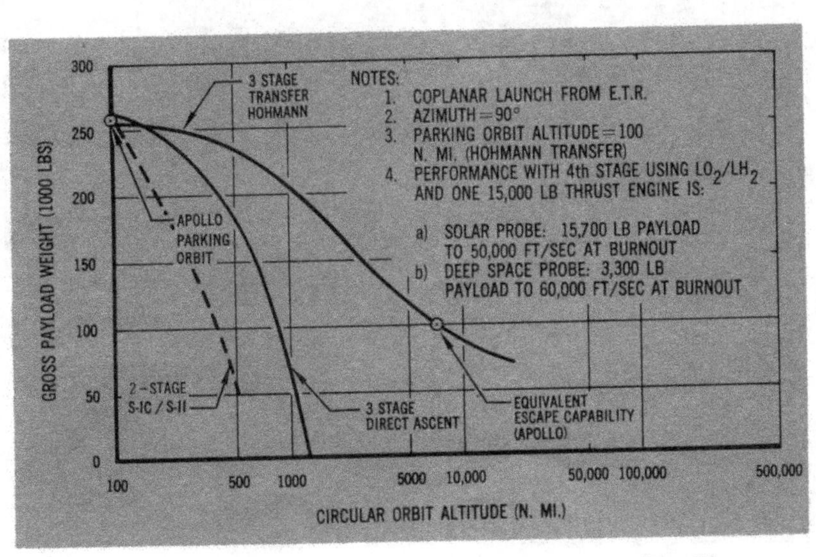

Figure I-2
SATURN V MISSION POTENTIAL

Figure II-1
SATURN V PAYLOAD POTENTIAL

3-STAGE AUXILIARY PAYLOADS

AREA		VOLUME (FT^3)	WEIGHT (LBS)	EXPERIMENT CONTACT AGENCIES (5)
COMMAND (1) MODULE	BLOCK I	7.2	< 80	NASA-MSC/NAA-S&ID
	BLOCK II	3	< 80	
SERVICE MODULE (1)		(3)	(3)	NASA-MSC/NAA-S&ID
LEM ASCENT (1)		3	< 80	NASA-MSC GRUMMAN
LEM DESCENT		15	210	NASA-MSC GRUMMAN
LEM ADAPTER		UP TO 3230	29,500 (2)	NASA-MSC/NAA-S&ID
INSTRUMENT UNIT	COLD PLATES	37	2400	NASA-MSFC/IBM/DAC
	CENTER	380	2000	
SATURN V/IVB	VOL. NO.			NASA MSFC NASA-WASH DC DOUGLAS AIRCRAFT CO
	1a	78	1100	
	1b	78	1100	
	2	100	1000	
	3	39	900	
	4a	45	500	
	4b	(SEE IU)	—	
	5	< 8 TOTAL	—	
	6	< 5 TOTAL	—	
	7	—	—	

PRIME PAYLOAD CAPABILITY

VEHICLE CONFIGURATION	GROSS PAYLOAD VOLUME (FT^3) (4)	GROSS PAYLOAD WEIGHT (LBS) (4)	
3 STAGE 100 N. MI.	TO 5000	261,000	NASA-MSFC-MSC NASA-WASH-DOUGLAS
3 STAGE 500 N. MI.	TO 5000	172,000	
3 STAGE ESCAPE	2990	98,000	
3 STAGE SYNCHRONOUS ALTITUDE	2990	72,000 ($i=28.5°$) 62,000 ($i=0°$)	

(1) NPC 500-9 APOLLO IN-FLIGHT EXPERIMENT GUIDE DATED SEPT. 15, 1964
(2) EQUAL TO TOTAL LEM WEIGHT INCLUDING PROPELLANTS
(3) SEE CONTACT AGENCIES
(4) FINAL AUXILIARY PAYLOAD WEIGHT AND VOLUME DEPENDS ON PRIME MISSION
(5) INFORMATION ON EXPERIMENT SUBMITTAL PROCESS AND ASSOCIATED VEHICLE DATA CAN BE OBTAINED FROM COGNIZANT NASA AGENCIES.

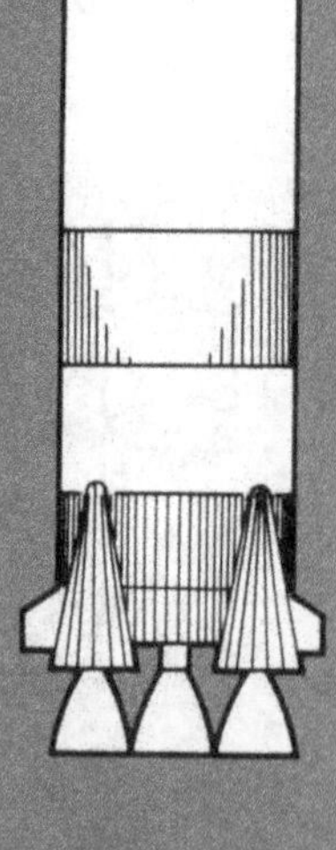

II-1. Planning and Schedules

Technical assistance is available at Douglas to aid the experimenter or payload originator in planning, flying and evaluating a payload on the Saturn V. The three-stage Saturn V vehicles can carry prime or auxiliary payloads on a great variety of manned or unmanned missions. Since it is beyond the scope of this guide to include all the data on each payload volume, some of the significant examples are shown in Figure II-1. Space, power and weight carrying capability is available in almost every part of the vehicle. Depending on specific mission requirements, auxiliary payloads may be carried in the:

(a) Apollo Command Module

(b) Apollo Service Module

(c) Lunar Excursion Module (LEM) Ascent or Descent Modules

(d) LEM Adapter

(e) Instrument Unit (I.U.)

(f) S-IVB Stage

(g) Fourth Stage (in four-stage, high energy mission configurations)

A summary of Saturn V payload potentials is presented in Figure II-1. New prime payloads may use either existing or special shroud designs.

Information in this guide is primarily associated with the prime payload carrying ability of the Saturn vehicle, auxiliary payloads within the S-IVB and payloads supported by the S-IVB and extending above it. Experimenters desiring more information on the other stages or modules should contact the appropriate agency as listed in Figure II-1.

The general steps normally required to bring a prime or auxiliary payload from concept, through integration and flight with the launch vehicle system, to final evaluation, are presented in the flow-diagram shown in Figure II-2.

Payloads and experiments may be conceived by any organization or individual in the government, universities or industry. In some cases, in order to be effective, payload proposals must include certain launch vehicle and program interface data. Douglas will assist experiment originators in the definition of payload/launch vehicle concepts. The payload/experiment proposals submitted by the originating organization are evaluated by NASA experiment review boards to determine the concept's priority in meeting national objectives. With mission objectives approved, budget and vehicle allocations can be made and the concept can be processed through normal procurement channels to obtain the final contractual authority.

Upon receipt of payload contractual authority, more detailed mission planning will be accomplished among NASA, the Saturn V stage contractors, and the payload originator. NASA acts as overall program integration manager.

Development and qualification of payloads proceed in parallel with launch vehicle production. Peculiar payload requirements may necessitate accomplishment of detailed testing, test support planning, and test documentation. These must be accomplished at the beginning of final checkout of the Saturn vehicle to ensure compatibility of payloads and launch vehicle.

PAYLOAD CONSIDERATION

Figure II-2
PAYLOAD PLANNING AND IMPLEMENTATION FLOW DIAGRAM

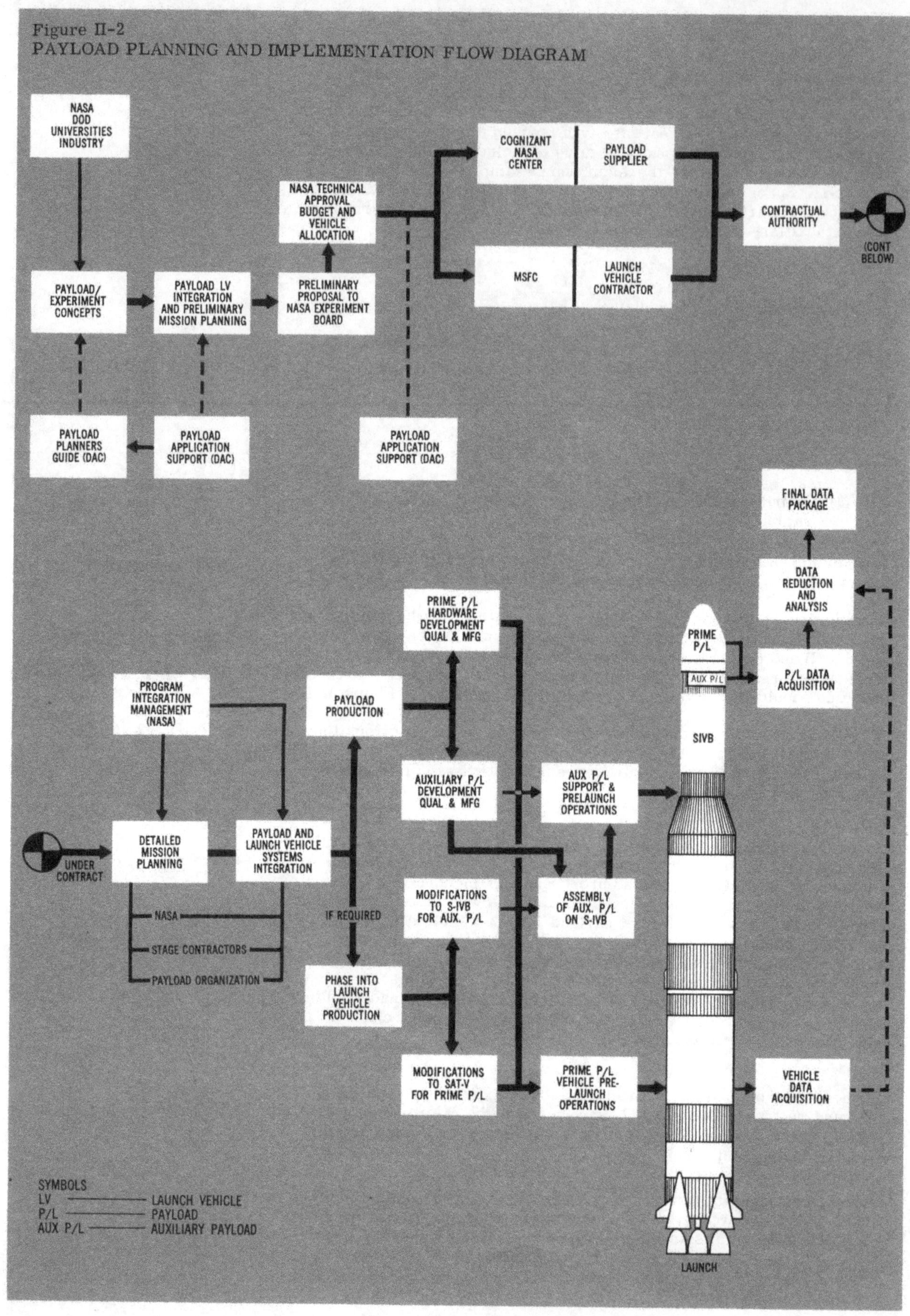

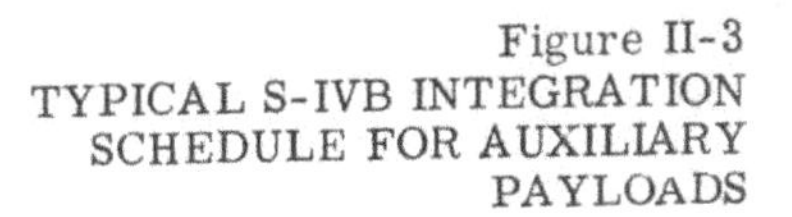
Figure II-3
TYPICAL S-IVB INTEGRATION SCHEDULE FOR AUXILIARY PAYLOADS

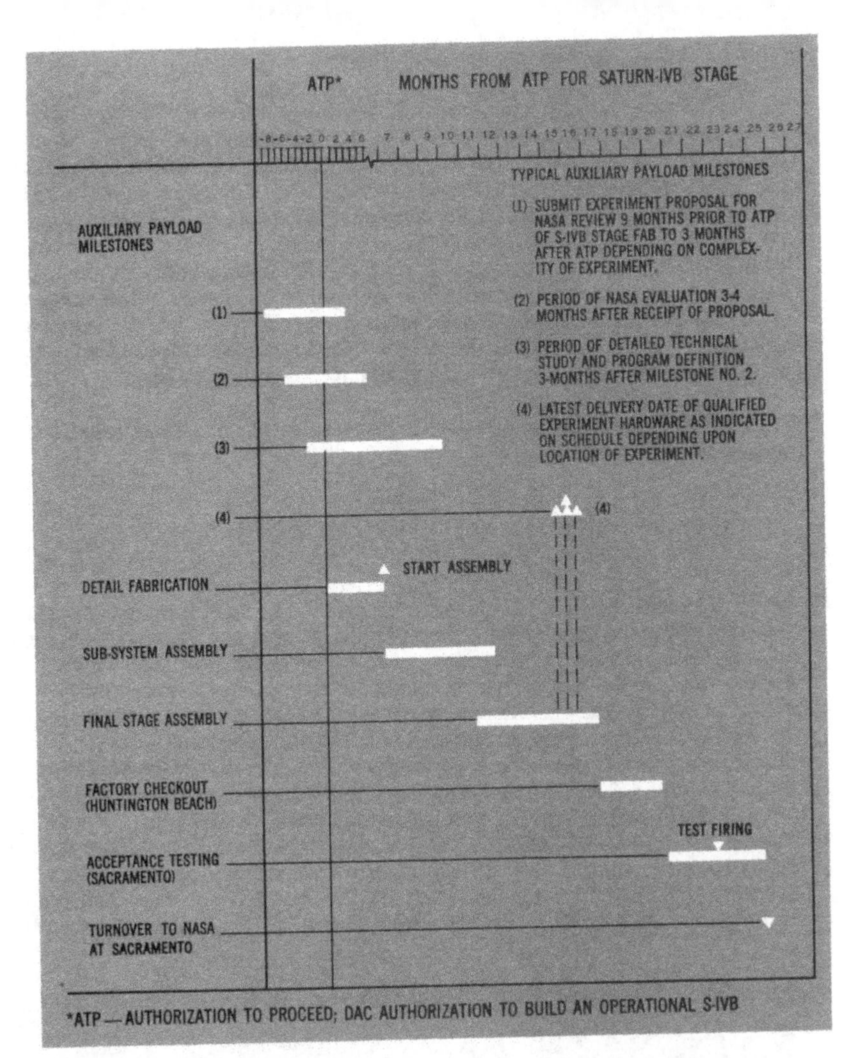

A typical schedule of S-IVB stage production and critical auxiliary payload integration periods is shown in Figure II-3. Also shown are typical delivery dates that an S-IVB mounted auxiliary payload might have to meet to minimize interference with the delivery schedule of the stages. The complexity of the payload and the nature of its integration will establish the lead time for a particular flight.

A schedule indicating the type of operations that must be accomplished at Kennedy Space Center (KSC) to prepare a prime payload is presented in Figure II-4.

Figure II-5 indicates typical delivery dates to KSC for Saturn V vehicles SA-501 through SA-515. Deliveries beginning with SA-516 may be estimated at a rate of six per year. While most of these currently have prime payload assignments, some are not expected to be fully loaded and may have room for auxiliary payloads.

II-2. Launch Vehicle Accommodations

Since auxiliary payloads can vary widely in size, shape and weight, the S-IVB stage has been reviewed in detail to identify locations in which auxiliary payloads can be carried if a weight allowance is available on a flight. Several volumes may be used depending upon the experimenter's specific requirements.

II-2-1. Auxiliary Payloads Mounted in S-IVB Stage

Convenient volumes may be available to experimenters in the forward portion of the stage and in the I.U. The envelopes of available space within the forward skirt and I.U. extend from the electrical/electronic units mounted on the skirt to the forward dome of the S-IVB tank as shown in Figure II-6. Also, additional space is available in pods mounted externally on the forward skirt. Many combinations of space, power, data, and environmental systems can be furnished to meet the needs of auxiliary payloads. These systems do not exist in the present vehicle. This discussion is intended to illustrate feasible techniques which could be employed to accommodate auxiliary payloads.

The possible experimental payload volumes within the Saturn S-IVB stage are listed below:

a. Experiment Volumes No. 1a and 1b (Figure II-6)

About 78 cubic feet could be provided external to the forward interstage in each of two pods as shown. Since these pods have not been designed, it may be possible to include provisions for certain unique payload requirements in the basic layout of these volumes. Approximately 1,100 lb of payload may be carried in each location. Some modification to the forward skirt for structural support and rerouting of some electrical cables will be required. Mounting concepts for these experimental volumes are shown in Figure II-7. The first concept shows a payload incorporating a solid propellant motor (ABL-X-258), an optional second motor (ARC-XM-85), and a satellite payload. The assembly is mounted in a support cradle that also provides means of ejection from the S-IVB stage. The ABL-X-258 motor is ignited by a signal from a timer keyed to the ejection sequence. The payload assembly

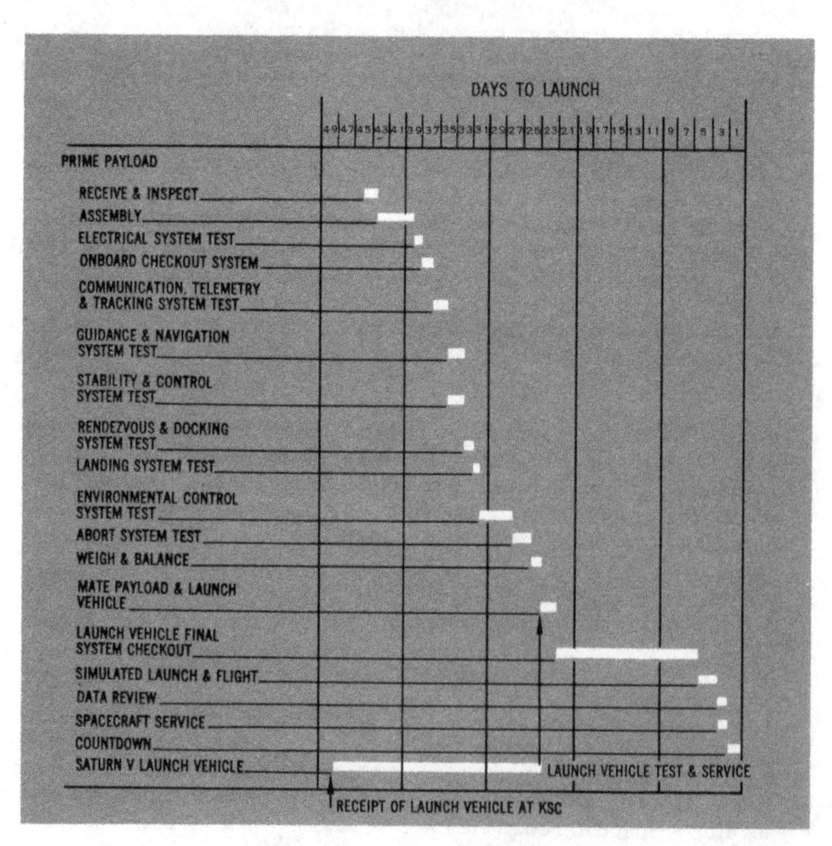

Figure II-4
TYPICAL KSC SATURN PREPARATION SCHEDULE FOR PRIME PAYLOAD

has an attitude control system referenced to the S-IVB stage attitude at payload separation. The payload is protected through the boost phase of the trajectory by a fairing that is jettisoned just before payload ejection. The support cradle is attached to a honeycomb mounting plate that in turn is attached to the S-IVB stage structure through the forward skirt frames and to pads on the S-IVB Liquid Hydrogen (LH_2) tank skin. Some performance figures for this type of installation are shown in Configuration B Figure II-8. Other payloads with different requirements that could also be accommodated are indicated by concepts 2, 3, and 4 of Figure II-7.

b. Experiment Volume No. 2

Variation in the shape of Volume 2 is possible depending on the payload configuration. However, some limitations on the use of this space are set by checkout requirements on equipment and wiring in the interstage, I.U., and the LEM descent module. Accessibility to these areas requires the use of a vertical access kit that restricts the available volume to that under the access kit platform. This volume consists of approximately 109 cubic feet. The experiment modules can be mounted on a lightweight structural cone supported by one of the forward skirt frames. A total payload weight of about 1,000 lb can be carried in this location. Weight limitations on a specific experiment module must be controlled by prime mission requirements as well as by structural design factors.

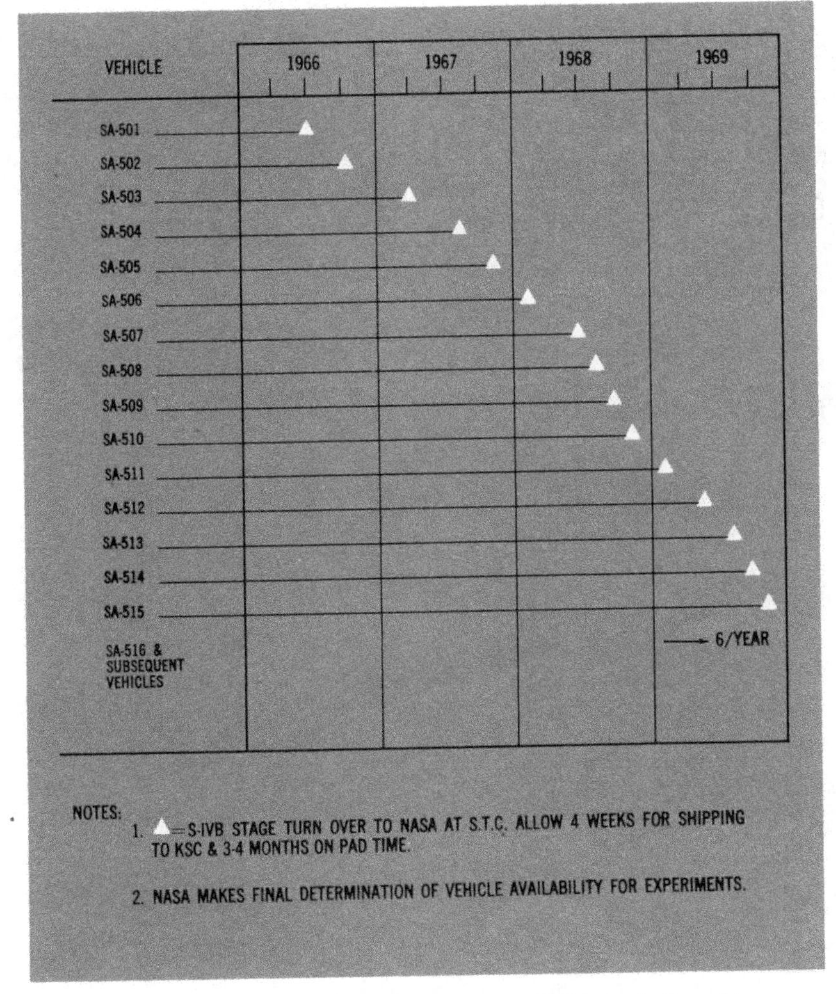

Figure II-5
SATURN S-IVB
DELIVERY SCHEDULE

Figure II-6
PROPOSED CONCEPTS FOR S-IVB AUXILIARY PAYLOAD VOLUMES

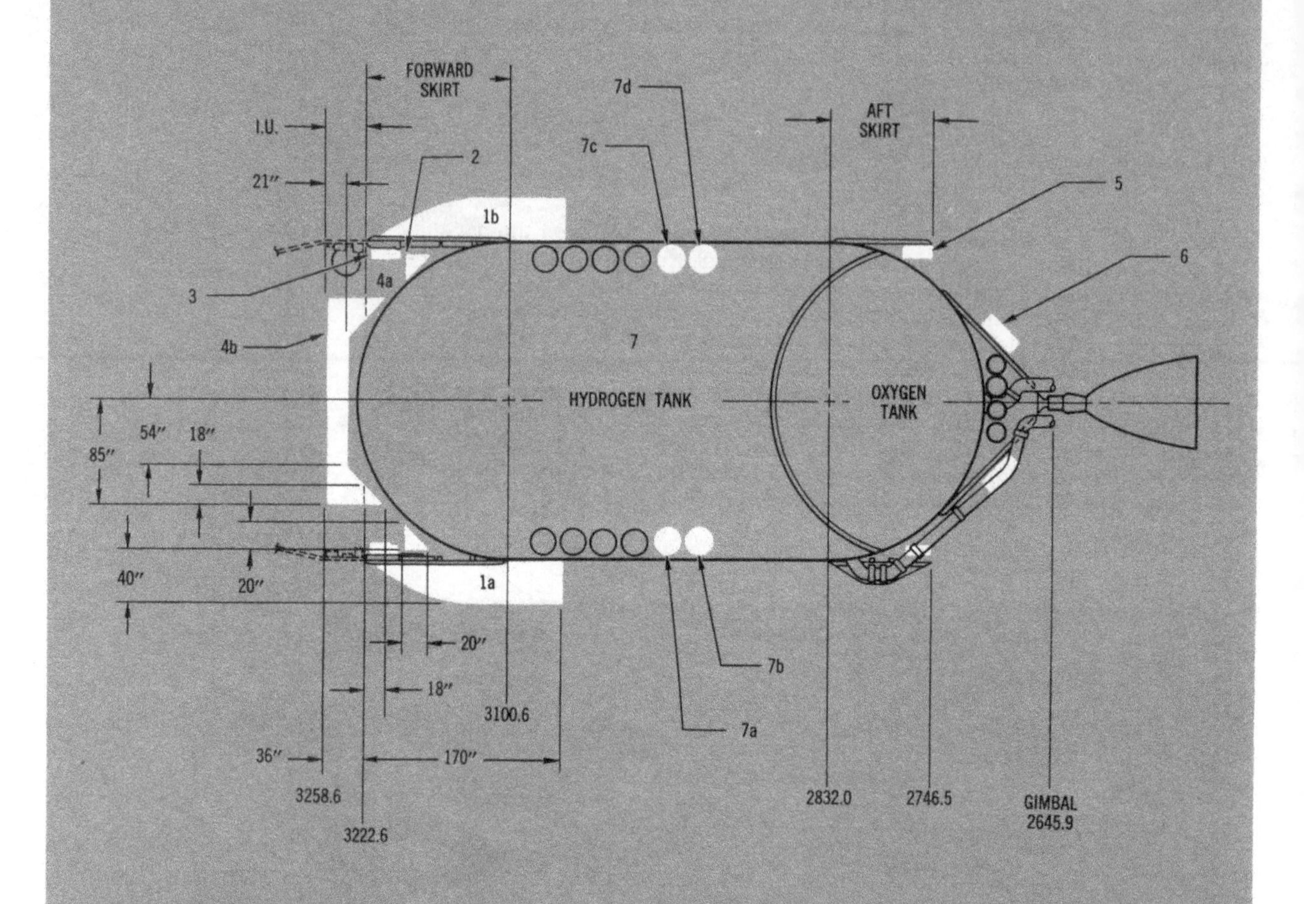

VOL. NO.	LOCATION	VOLUME	PAYLOAD WEIGHT, LBS
1a	FWD. SKIRT — EXT.	78 FT^3	1,100
1b	FWD. SKIRT — EXT.	78 FT^3	1,100
2	FWD. SKIRT — INT.	109 FT^3	1,000
3	FWD. SKIRT — INT.	39 FT^3	900
4a	FWD. SKIRT — INT.	45 FT^3	2,500
4b	I.U. — INT.	380 FT^3	
5	AFT SKIRT — INT.	8 FT^3	—
6	THRUST STRUCTURE	INDEF.	—
7	HYDROGEN TANK	—	—
7a-d	HYDROGEN TANK	3.5 FT^3 EA.	—

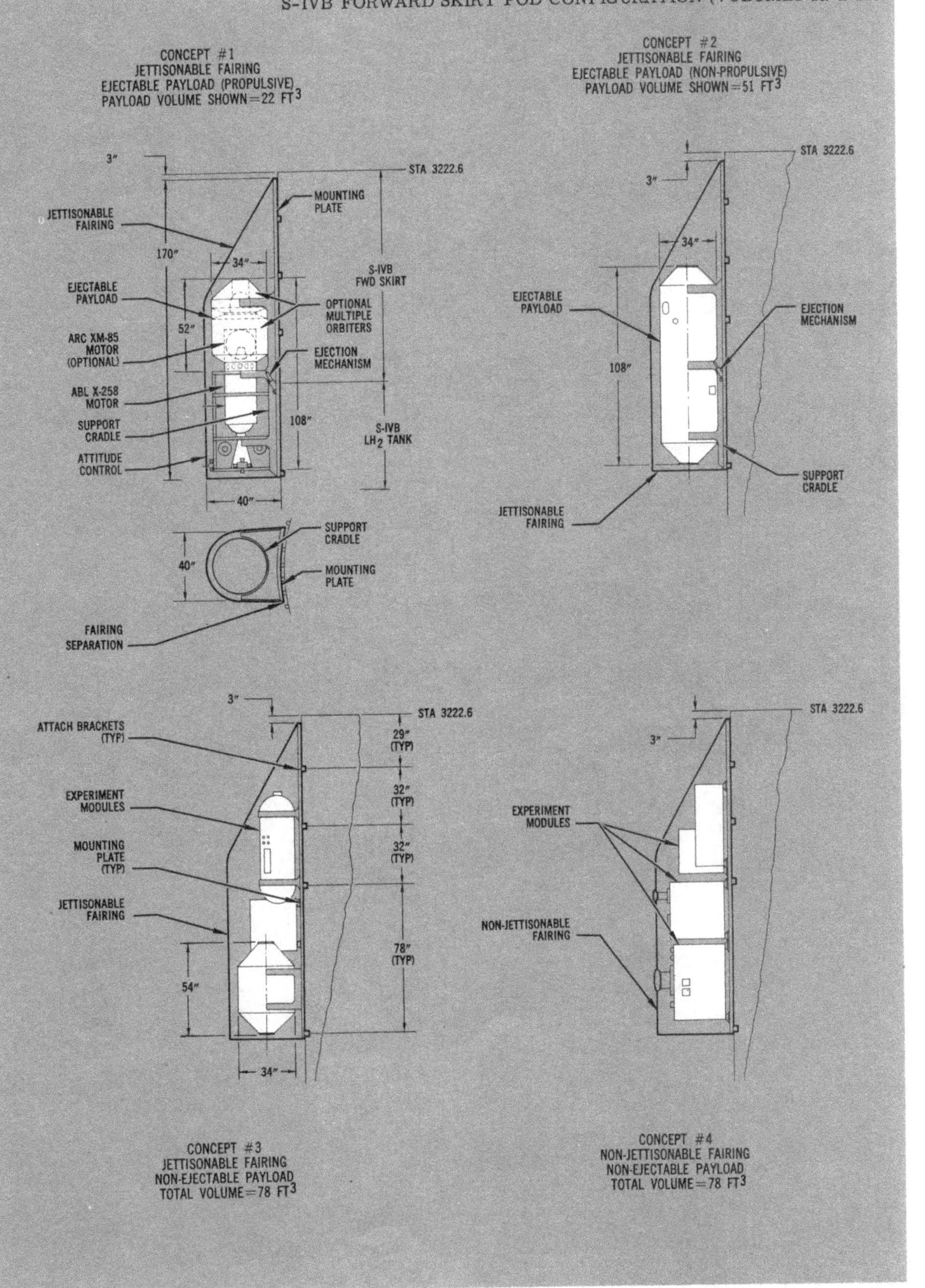
Figure II-7
S-IVB FORWARD SKIRT POD CONFIGURATION (VOLUMES 1a & 1b)
CONCEPT #1
JETTISONABLE FAIRING
EJECTABLE PAYLOAD (PROPULSIVE)
PAYLOAD VOLUME SHOWN=22 FT3
CONCEPT #2
JETTISONABLE FAIRING
EJECTABLE PAYLOAD (NON-PROPULSIVE)
PAYLOAD VOLUME SHOWN=51 FT3
3"
STA 3222.6
MOUNTING PLATE
JETTISONABLE FAIRING
170"
34"
S-IVB FWD SKIRT
EJECTABLE PAYLOAD
OPTIONAL MULTIPLE ORBITERS
52"
ARC XM-85 MOTOR (OPTIONAL)
EJECTION MECHANISM
ABL X-258 MOTOR
SUPPORT CRADLE
108"
S-IVB LH2 TANK
ATTITUDE CONTROL
40"
SUPPORT CRADLE
MOUNTING PLATE
FAIRING SEPARATION
EJECTION MECHANISM
SUPPORT CRADLE
ATTACH BRACKETS (TYP)
29" (TYP)
32" (TYP)
32" (TYP)
78" (TYP)
EXPERIMENT MODULES
MOUNTING PLATE (TYP)
54"
NON-JETTISONABLE FAIRING
CONCEPT #3
JETTISONABLE FAIRING
NON-EJECTABLE PAYLOAD
TOTAL VOLUME=78 FT3
CONCEPT #4
NON-JETTISONABLE FAIRING
NON-EJECTABLE PAYLOAD
TOTAL VOLUME=78 FT3

Figure II-8
SATURN S-IVB STAGE ALTERNATE CONFIGURATIONS FOR AUXILIARY PAYLOADS

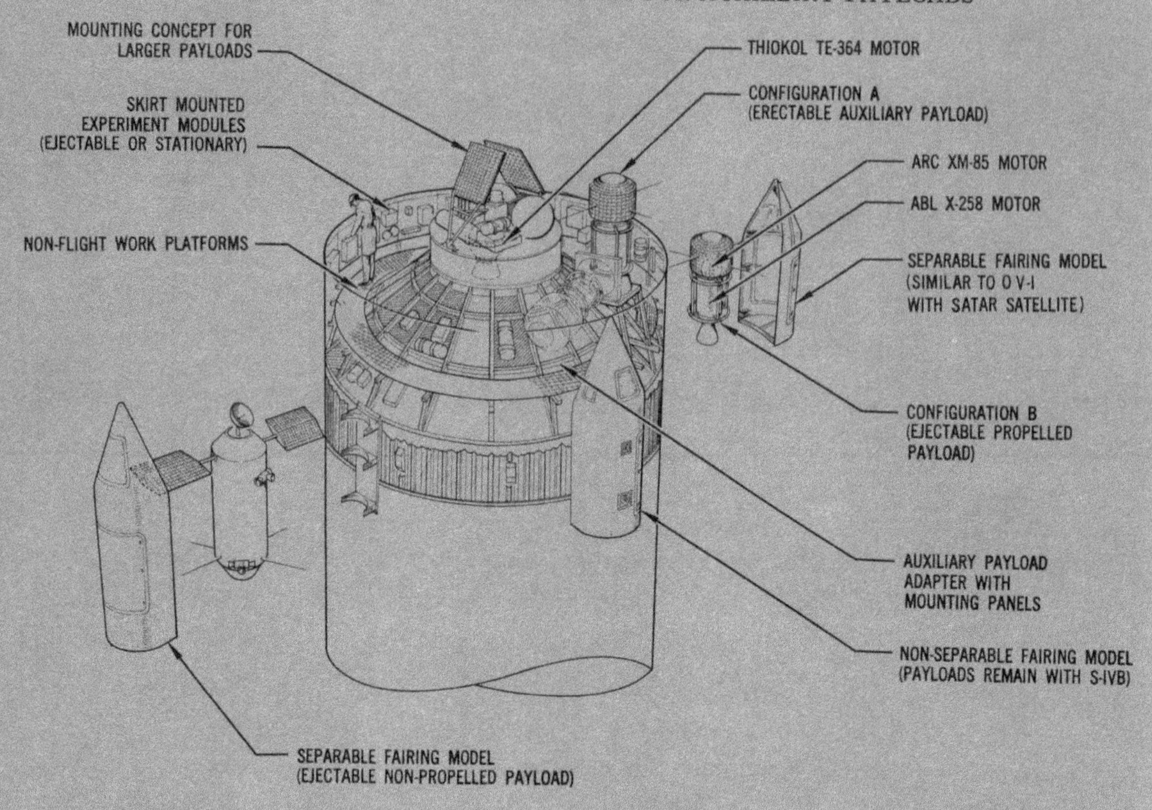

CONFIGURATION A
(ABL X-258 MOTOR + SATELLITE)

ELLIPTICAL ORBIT CAPABILITIES
(INITIAL ORBIT ALTITUDE=100 N. MILES)

SATELLITE WEIGHT (LBS)*	APOGEE (N. MI.)	PERIGEE (N. MI.)
157	LUNAR MISSION	—
200	44,000	100
300	15,300	100
400	9,500	100
500	7,000	100

*SATELLITE WEIGHT INCLUDES GUIDANCE SYSTEM

CONFIGURATION B
(ABL X-258 MOTOR + ARC XM-85 MOTOR + SATELLITE)

CIRCULAR ORBIT CAPABILITIES**
(INITIAL ORBIT ALTITUDE=100 N. MI.)

SATELLITE WEIGHT (LBS)*	ORBITAL ALTITUDE (N. MI.)
150	7,700
200	4,700
300	2,850

ELLIPTICAL ORBIT CAPABILITIES
(INITIAL ORBIT ALTITUDE=100 N. MI.)

SATELLITE WEIGHT (LBS)*	APOGEE (N. MI.)	PERIGEE (N. MI.)
100	24,700	17,000
200	9,800	6,600
300	7,000	3,500

**REQUIRES OFF-LOADING X-258 MOTOR

c. Experiment Volume No. 3

Experimental modules can be mounted directly on the thermal conditioning panels in the forward skirt. See Figure II-9. On all vehicles beginning with SA-504, six or more of the sixteen panels will be available for mounting experiments because of a simplified telemetry system. A volume of at least 39 cubic feet with a maximum weight of 900 lb is available. This weight and volume indicated may be increased if the accessibility, the payload center of gravity location and the mounting method permit.

d. Experiment Volume No. 4

For some missions in which the LEM descent stage is not carried, an additional large volume may be available above the access kit platform. This space extends over the S-IVB forward dome and into the instrument unit to Station 3258.6. The volume available within the forward skirt is about 45 cubic feet. Approximately 380 cubic feet is available in the I.U. The experiment modules can be mounted on an auxiliary payload adapter. This payload adapter consists of a 'spider' structure supported from the S-IVB forward skirt frames as shown in Figure II-10. The adapter would also serve as an access kit when removable work platforms are inserted as shown. The experimental modules are mounted on honeycomb panels attached to the adapter. The adapter accompanying the modules must be removable for access to the liquid hydrogen tank. A total payload weight of up to 2,500 lb may be carried in this location, prime payload weight permitting.

Other auxiliary payloads, such as the Delta third stage, may be carried as shown in Figure II-8. The payloads are mounted on the auxiliary payload adapter through additional supporting structure. The internal mounting depicted shows the Delta third stage including separation and spin-up mechanism. The payload is carried in a horizontal or stowed position during the boost phase until the separation of the S-IVB from the prime payload. At this time the Delta third stage is erected, spun-up, and separated at a signal in the S-IVB stage separation sequence. Ignition of the ABL-X-258 motor is triggered by a timer after an appropriate separation distance is achieved. Separation forces can

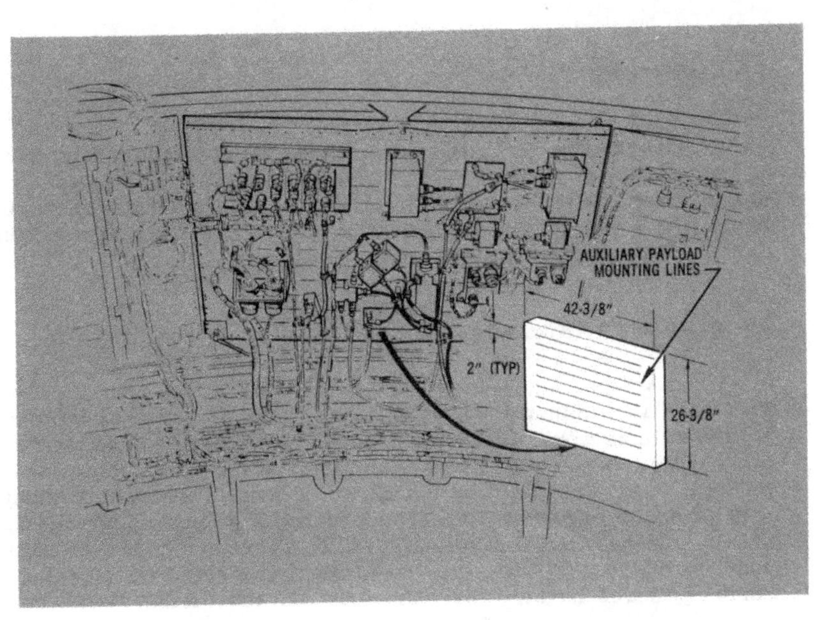

Figure II-9
S-IVB FORWARD
SKIRT THERMAL
CONDITIONED PANELS

Figure II-10
AUXILIARY PAYLOAD ADAPTER

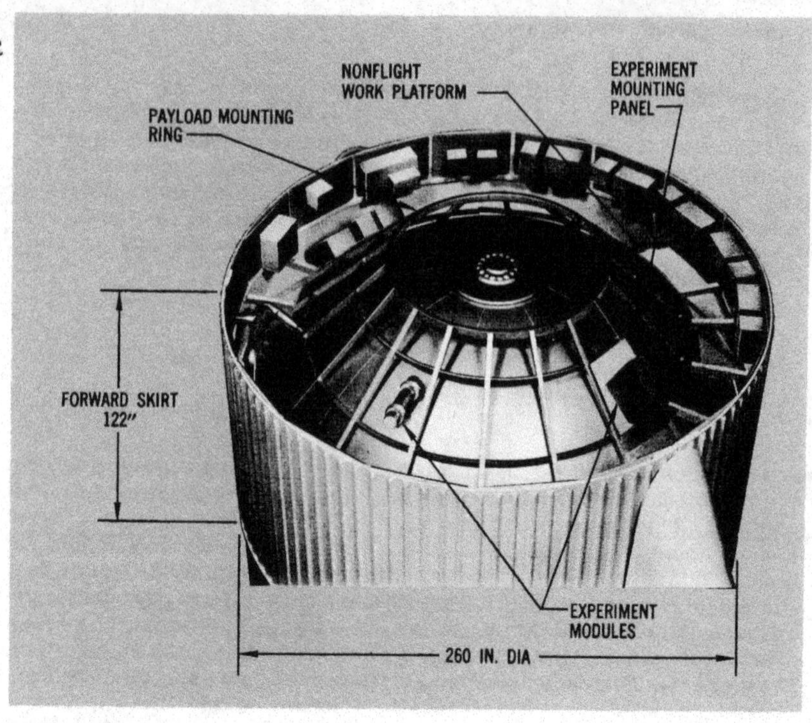

be generated by small solid propellant motors similar to the spin rockets. In the stowed position the Delta third stage projects approximately 33 inches above the S-IVB/I.U. interface at Station 3222.6. Some representative performance figures for two possible configurations are shown (Figure II-8).

e. Experiment Volume No. 5

A small amount of usable volume may be available in the aft skirt area of the operational configuration. Certain modules (five volumes of about 1.5 cubic feet each) may be mounted directly on the existing mounting plates in place of R&D equipment not required on operational flights.

f. Experiment Volume No. 6

Experimental modules of light weight may be mounted directly on the thrust structure. Precise locations and volumes available cannot be defined at present, but small modules of the proper size (about one cubic foot each), shape, and weight could be accommodated depending on the mounting requirements of the payloads.

g. Experiment Volume No. 7

Volume 7 is within the hydrogen tank itself. Any experiment placed in this volume would, of course, displace the LH_2 and be subjected to the temperature and pressure conditions of the LH_2. Some experimenters may want to take advantage of these conditions to study a system under cryogenic and space environment, or to study the fluid behavior of the liquid or gaseous hydrogen. There are eight cold helium spheres in the LH_2 tank to pressurize the liquid oxygen (LOX) tank during powered flight. There are four additional flanged connections on which spheres could be installed to hold experiments at liquid hydrogen temperatures while protecting them from direct contact with the hydrogen. Each of the spheres has a volume of 3.5 cubic feet. The entrance to the sphere is only 1.44 inches in diameter. However, this could be increased to about 4 inches in diameter.

II-2-2. Prime Payloads Above the S-IVB Stage

Minimum vehicle changes will be required if the volume normally occupied by the Lunar Excursion Module (LEM) were to be utilized for other payloads (Figure II-11). This volume within the LEM adapter might be used on future flights if prime mission objectives permit.

Figures II-12, II-13 and II-14 illustrate three other possible configurations of payloads and payload fairings. The fairings protect the payloads from aerodynamic loads and temperatures while in flight. They may also be used for payload thermal conditioning on the launch pad, if such conditioning is required and provisions for it are included. The fairings may be made of aluminum honeycomb or of fiberglas, if RF transparency is a requirement. The fairings are jettisoned during second-stage operation when atmospheric effects are negligible.

A typical adapter which supports the prime payload, Figure II-15, can be designed to the diameter dictated by payload requirements. The adapter is mounted directly on the instrument unit and is a conical frustrum of semi-monocoque construction. The adapter includes a structural ring to bear the lateral components of the loads imposed by the payload and to provide clearance for the end frame of the fairing at Station 3264.6. The height of the adapter above the mating plane at Station 3264.6 is shown as 36 inches. This dimension can be varied, if required. The weight would depend on the prime payload weight and must be accounted for when estimating vehicle performance.

Configuration "A" shown in Figure II-11 utilizes the volume that would be available if mission objectives are such that the LEM is not used. A payload volume of about 3,230 cubic feet and a weight of 29,500 lb could be accommodated in this space. The payload could remain with the S-IVB in orbit, or be ejected after separation of the forward section of the spacecraft/LEM adapter. The LEM adapter incorporates four panels that are unfolded at the time of command module separation.

Configuration "B" shown in Figure II-12 is the shape originally designed for the Voyager spacecraft and encompasses a volume of approximately 2,990 cubic feet. These dimensions are approximate and should be used for preliminary layout only. The final dimensions depend on payload configuration and adapter height requirements. The approximate weight of the fairing is 2,500 lb if made of aluminum honeycomb.

Configuration "C" shown in Figure II-13 is for a modified LEM adapter and encompasses a volume of about 5,000 cubic feet with the approximate dimensions shown. The weight of the fairing is about 3,000 lb.

Configuration "D" shown in Figure II-14 combines the Voyager nose fairing with an S-IVB stage forward skirt. It encompasses a usable volume of approximately 6,000 cubic feet and weighs about 3,600 lb.

a. Prime Payload Attitude Control Systems

Prime payloads may require their own attitude control systems if they are to be separated from the S-IVB stage during orbital coast.

A concept using existing Saturn IB/S-IVB(A) or Saturn V/S-IVB(B) Auxiliary Propulsion System (APS) modules on prime payloads is presented in Figure II-16. The APS modules are presently designed for 4-1/2 and 6-1/2 hours coast, respectively, when mounted on the

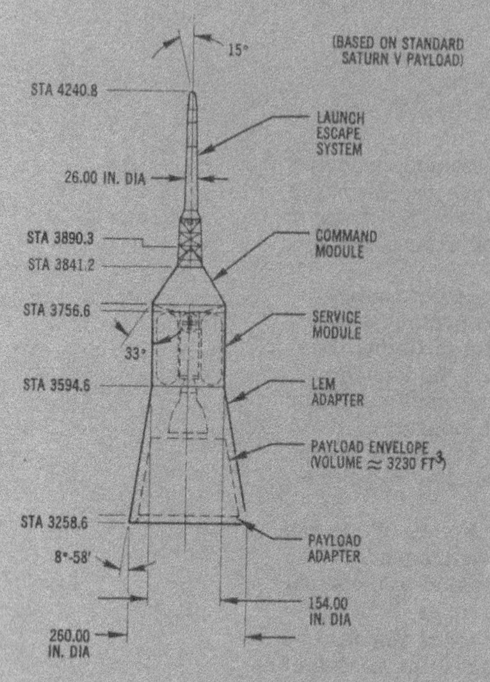

Figure II-11
PRIME PAYLOAD FAIRING (CONFIGURATION A)

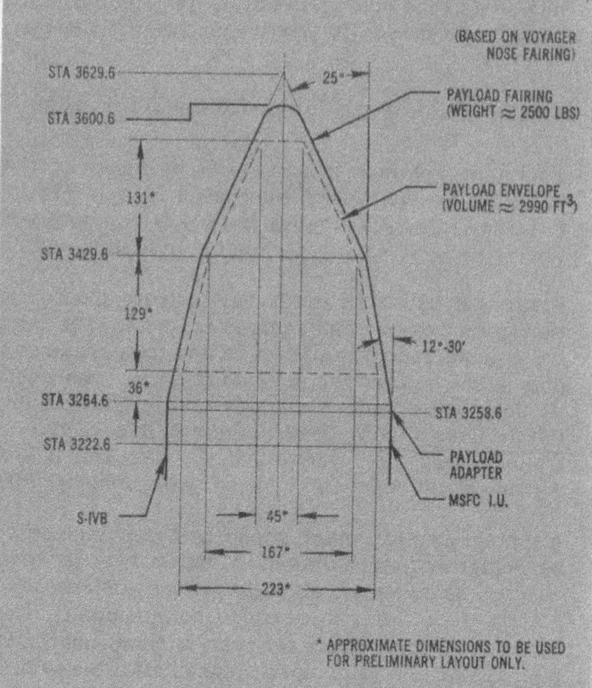

Figure II-12
PRIME PAYLOAD FAIRING (CONFIGURATION B)

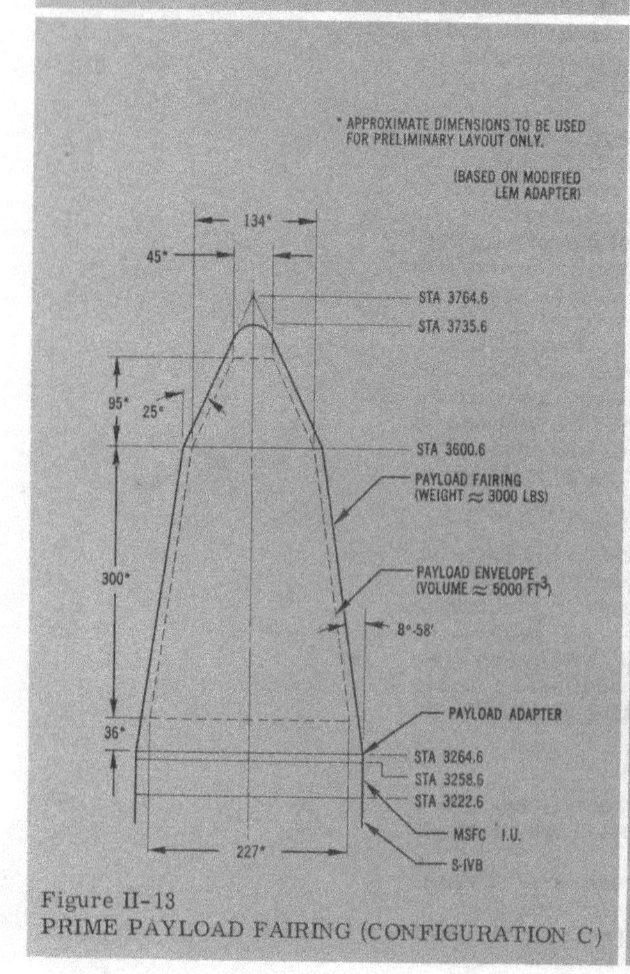

Figure II-13
PRIME PAYLOAD FAIRING (CONFIGURATION C)

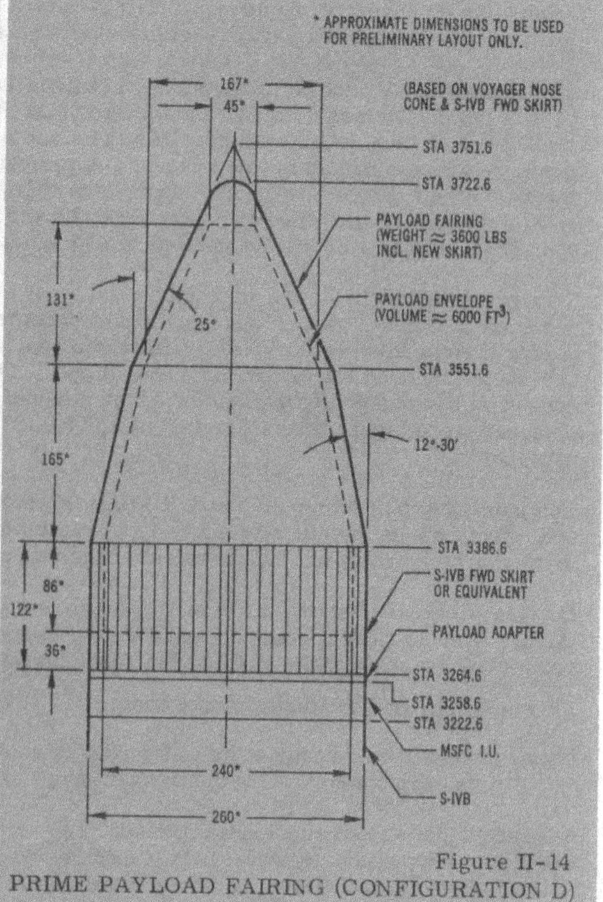

Figure II-14
PRIME PAYLOAD FAIRING (CONFIGURATION D)

Figure II-15
PRIME PAYLOAD ADAPTER

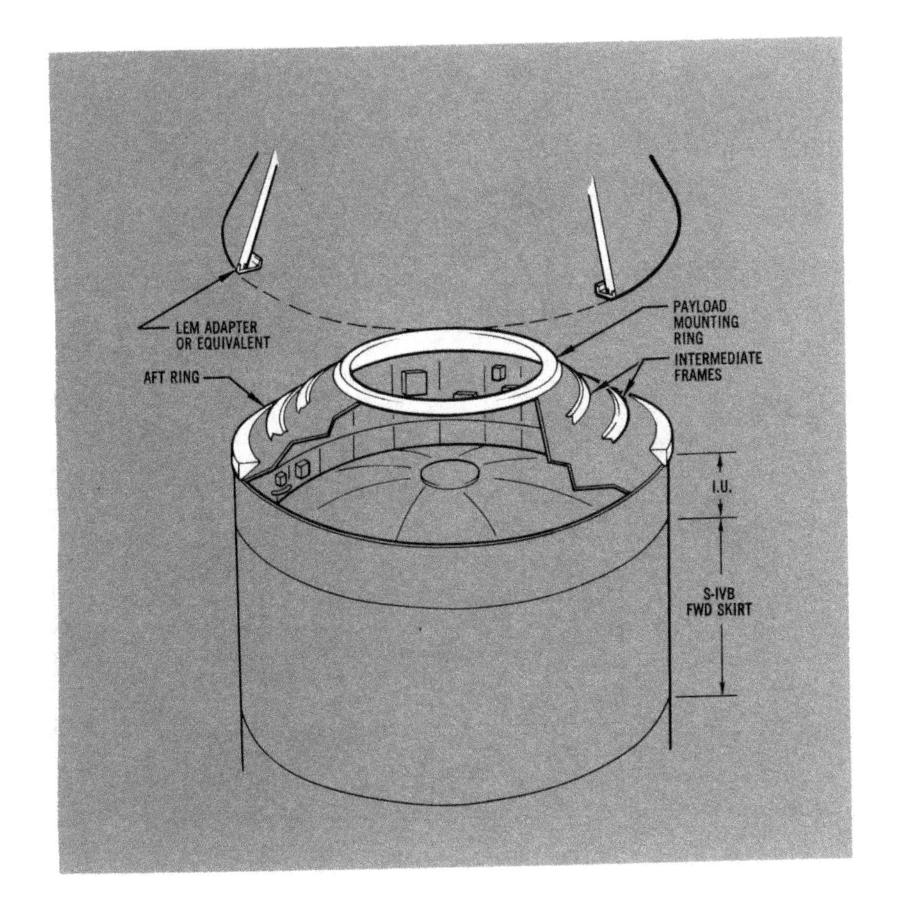

S-IVB stage used in the Saturn IB and V missions. Much longer coast periods can be achieved if these units are used for payload attitude control. The duration will be a function of payload moment of inertia and required operating cycle.

The APS modules are self-contained propulsion units which require electrical power, vehicle attitude sensors, control circuitry and guidance signals. The guidance and attitude sensing signals are provided by the I.U. The electrical power requirements for either the Saturn V or IB modules are 28 volts at a maximum of 26.5 amp for operating valves and switches. The attitude control band requirements of the payload, moments of inertia, center of gravity, location of the payload, and environmental disturbances dictate the total propellant needed for a given mission. The 150 lb thrust is, perhaps, larger than necessary but there are techniques available for reducing it by about 50% for better propellant economy. Smaller engines from other programs could also be used. However, with the payloads indicated in Figure II-16, for the Saturn V module, control periods in excess of 6-1/2 hours with a deadband of $\pm 1^{\circ}$ in all three axes are possible. Reducing the control accuracy requirements extends the operating duration. Detailed descriptions of the APS modules and techniques for extending their operational life are presented in Section II-5.

II-3. Payload Thermal Environment and Control

Accurate determination of the payload temperature control requirement demands realistic thermodynamic models. Douglas employs a series of heat transfer computer programs (both 1 and 3 dimensional) which indicate the temperature history that can be expected at any point in the vehicle for a multitude of thermodynamic environments. Should the temperature of the volume be critical for a particular experiment, a thermal protection system can then be designed.

Figure II-16
PRIME PAYLOAD USING S-IVB AUXILIARY PROPULSION SYSTEM

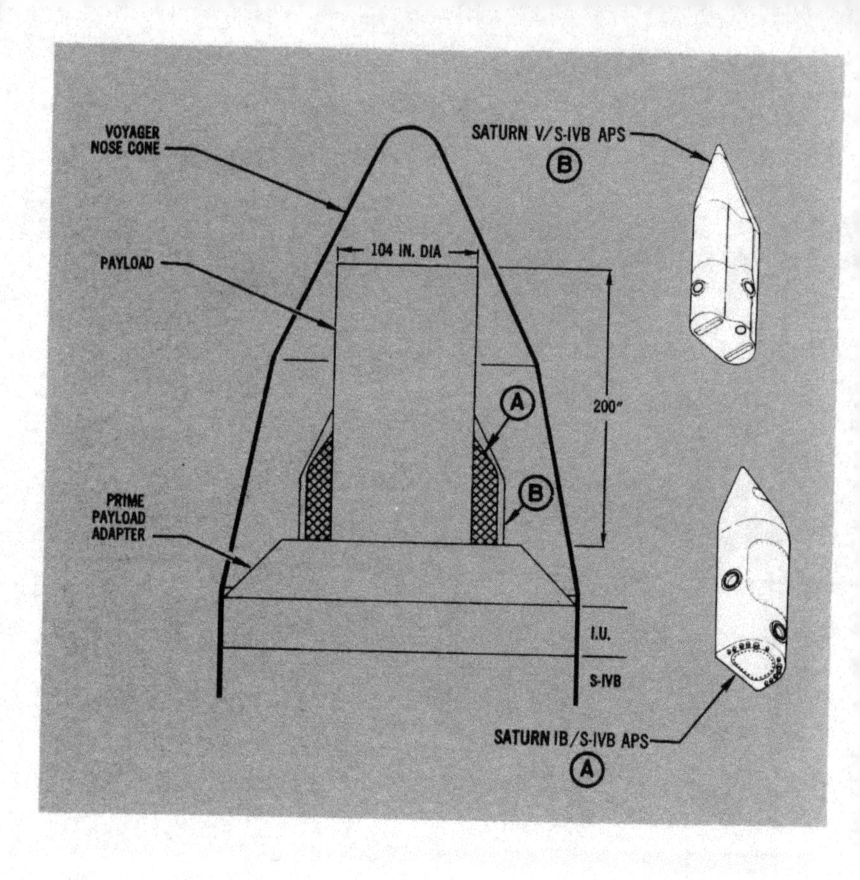

The forward skirt thermal control systems for cooling electronic equipment differ from those used in the aft skirt. The equipment mounted on the forward skirt is conditioned actively and that mounted aft is conditioned passively.

In the active system, electronic components are mounted on 16 thermal conditioned panels (cold plates, Figure II-9) which transfer heat to a coolant (60% methanol and 40% water) flowing through the panel. For the present S-IVB and I.U. flight plans, the coolant will enter the cold plates at 60°F maximum and leave at 70°F maximum. A coolant flow rate of approximately 0.5 gpm per panel is used at present. Units generating high heat loads should not be mounted close together since the coolant may not be capable of removing the required heat and excessive temperatures could result. The total allowable heat load per panel is 500 watts.

The mounting methods and the vibration levels predicted during launch allow 150 lb of equipment to be carried on each plate. Concentrated loads should be avoided. Experiments must be designed so that they can be mounted without interference to the coolant channels.

No cooling is available from the end of the prelaunch phase until approximately 130 sec after lift-off. A pre-launch purge gas system, utilizing air and gaseous nitrogen, provides the forward skirt area with a warming medium. It operates only up to the time of launch and provides no thermal control after that time. This system protects the electronic components and reduces oxygen present to 4% by volume. The total flow rate in the forward area is about 275 lb/min. The purge gas surrounding the components located in the I.U. and S-IVB forward skirt will be at a temperature of 35°F to 75°F.

In the passive system, electronic components are mounted in the aft skirt area on 18 fiberglass panels. No fluid thermo-conditioning system is used. Temperature is controlled through the proper surface finish of each electronic package and by providing conduction paths and insulation. Appropriate coatings are added when a special heating or cooling problem is revealed by calculation or test.

As now designed, each fiberglass panel in the aft skirt area is capable of supporting 100 lb of electronic packages. Four panels will be available for auxiliary payloads.

A separate pre-launch purge gas system maintains the equipment mounted on the aft skirt at a temperature of 20°F to 70°F during prelaunch procedures. Dry air at a flow of 300 lb per minute to the S-IVB stage is provided from a ground source. Gaseous nitrogen purge of approximately the same flow rate is initiated about 30 minutes before LH_2 loading. During flight, heat is radiated to space and to local sinks such as the LOX tank. If possible, high heat dissipating components or temperature-sensitive components should be mounted on the cold plates in the forward skirt.

If the above systems do not meet the needs of an experiment, modifications can be made to the thermal conditioning system or the purge gas system. Example of such changes are:

(a) Coolant flow rate in the thermal conditioning system can be changed to control the temperature of the experiment equipment.

(b) A space radiator could be installed to cool electronic equipment for long periods of time.

(c) Insulation and thermal control coatings may be engineered.

(d) Mounting procedures and requirements can be altered to vary heat conduction paths.

(e) Flow rate and temperature of the purge gas system could be varied.

(f) Purge gas could also be ducted directly to the experiment equipment.

II-4. Payload Acoustics and Vibration Environment

Acoustic and vibration phenomena have a similar time-history during a flight. A time-history of the former is shown in Figure II-17. The acoustic noise level inside the vehicle at three auxiliary and prime payload locations are shown in the figure. At lift-off, the exhaust of the first stage engines generates high frequency noise in the high shear mixing region close to the nozzles and lower frequency noise in the fully turbulent cores of the exhaust jets. This is transmitted through the air to the spacecraft and vehicle.

Following lift-off, the acoustic noise decreases as the exhaust pattern straightens out and as the distance between the vehicle and the ground reflecting surface increases. A further reduction occurs as the vehicle reaches supersonic speeds because the sound generated aft of the vehicle is left behind the vehicle. However, turbulent pressure fluctuations in the aerodynamic boundary layer intensify as free stream dynamic pressure increases. The maximum noise from this source occurs at the time of maximum dynamic pressure or shortly thereafter and it decreases as the dynamic pressure is reduced. The remaining excitation is structurally transmitted from engines, pumps, etc. These are of lower intensity and remain relatively constant until engine cut-off. Brief periods of vibration occur during retro rocket and ullage firings and stage separation.

To provide design criteria for payloads, design specifications have been developed which cover all of these environments. Figure II-18 is a broad-band acoustic specification for three payload locations, and represents an acoustical environment to which an item may be designed and ground tested to ensure satisfactory operation during an actual flight. All frequencies are assumed to be excited at the same time and at the appropriate level in each octave band. Figure II-19 is a broad-band random vibration specification for the same purpose. The duration of these qualification tests is longer than the duration of the significant environment for an actual flight to allow for exposure during static firings and to increase the reliability of the items. Figure II-20 is a sinusoidal vibration specification. The purpose of the sinusoidal sweep test requirement is to provide assurance that the item has adequate strength for transitory or unsteady phenomena that could occur in a flight. There is also a shock specification for each location but it is not included in this brief discussion.

II-5. S-IVB Stage Subsystem Information

There are at least four major subsystems of the S-IVB stage that may influence, or be of benefit to an auxiliary payload. They include the auxiliary propulsion, electrical power, thermal conditioning, and data acquisition systems.

II-5-1. Auxiliary Propulsion System

The Auxiliary Propulsion System (APS) modules for the S-IVB stage have two basic configurations as indicated in Figure II-21. The two are necessary to meet the mission requirements of the Saturn V vehicle and the Saturn IB vehicle. The major differences between the two are in propellant capacity and degrees of freedom.

The Saturn V/S-IVB APS is sized to provide roll control during powered flight, three axis attitude control during a 4-1/2 hour earth orbital coast and a two hour translunar coast, and propellant settling for continuous vent initiation and main engine restart. The attitude control function is provided by three 150 lb thrust engines in each module and propellant settling by a 72 lb thrust engine in each module. A mock-up of the Saturn V/S-IVB module is shown in Figure II-22. The Saturn IB/S-IVB APS does not provide for the translunar coast period nor does it have the 72 lb thrust engines.

The attitude control system is a pulse-modulated on-off system. The system is based on the minimum impulse capability of the 150-lb thrust engine, which has a minimum impulse bit capability of 7.5 lb-sec with an electrical input pulse width of approximately 65 milliseconds. The attitude control system is designed to operate with an attitude dead zone of ±1 degree in all axes. The undisturbed limit cycle rates of the Saturn V/S-IVB with payload in a 100 n. mi. circular orbit are approximately 0.02 deg/sec in roll and 0.001 deg/sec in pitch and yaw (0.003 deg/sec during translunar coast).

The auxiliary propulsion system is a completely self-contained modular propulsion sub-system. The modules require electrical power and command signals to provide the necessary stage functions. They are mounted on the aft skirt 180° apart. The equipment for loading propellants to the modules is a semi-automatic system, with individual umbilical connectors in each module.

Each module contains one 72 lb thrust and three 150-lb thrust ablatively-cooled liquid bi-propellant hypergolic engines, a positive expulsion (Teflon bladder) propellant feed system for zero gravity operations and a helium pressurization system. Each Saturn V APS mod-

Figure II-17
ACOUSTIC NOISE TIME-HISTORY

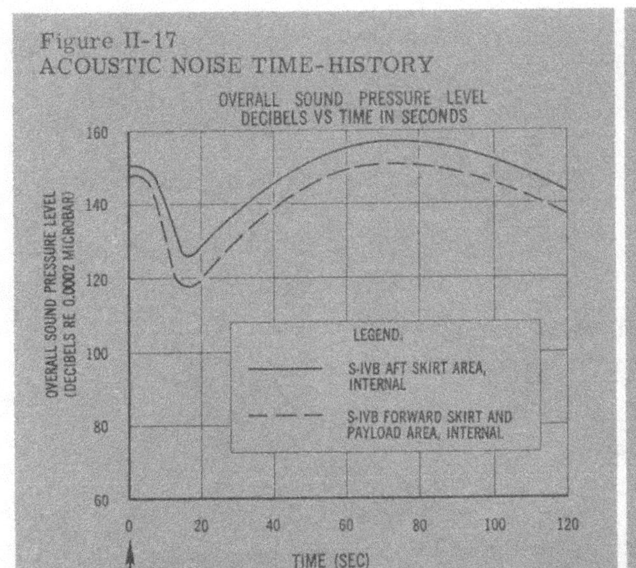

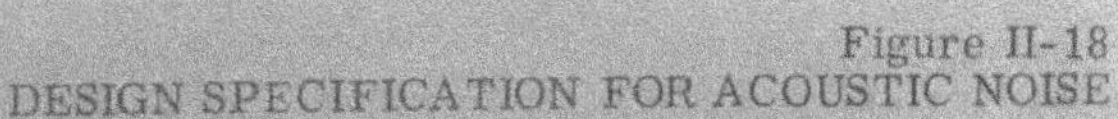

Figure II-18
DESIGN SPECIFICATION FOR ACOUSTIC NOISE

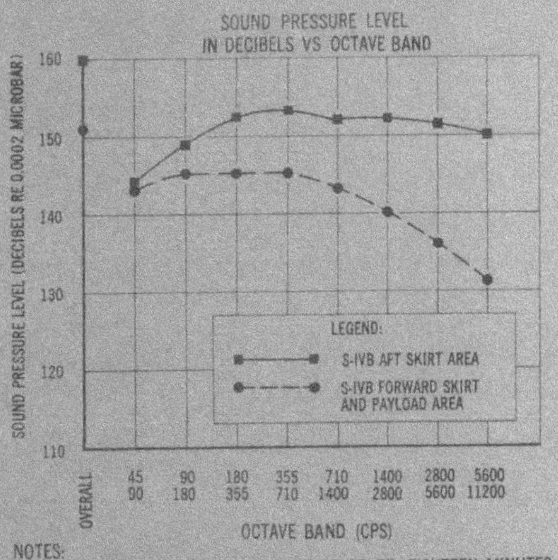

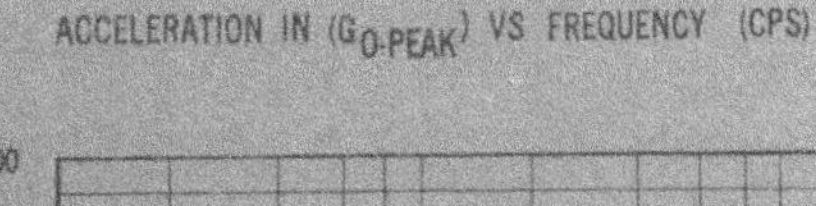

NOTES:
1. THE TIME DURATION IS ASSUMED TO BE EIGHTEEN MINUTES.
2. THE DIFFUSED SOUND FIELD OF RANDOM NOISE IS ASSUMED TO HAVE A GAUSSIAN AMPLITUDE DISTRIBUTION.

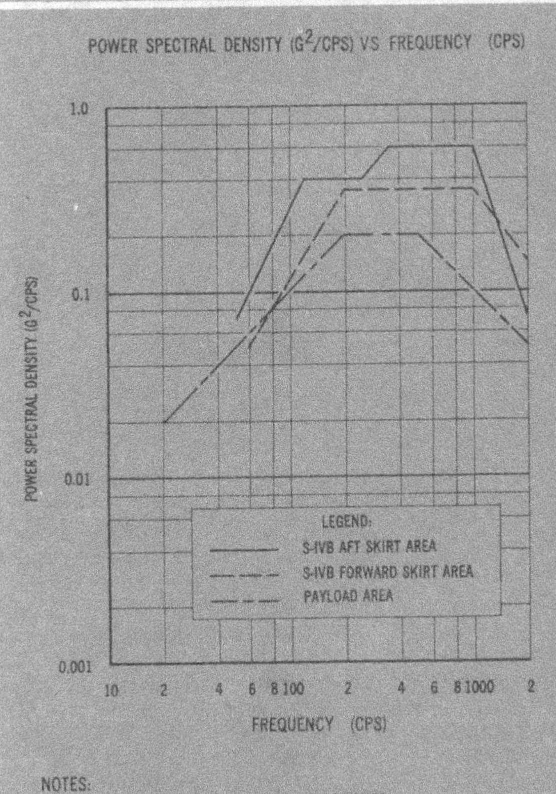

NOTES:
1. AMPLITUDE DISTRIBUTION IS ASSUMED GAUSSIAN
2. DURATION IS ASSUMED TO BE TWELVE MINUTES FOR EACH OF THREE MUTUALLY PERPENDICULAR DIRECTIONS.

Figure II-19
DESIGN SPECIFICATION FOR RANDOM VIBRATION

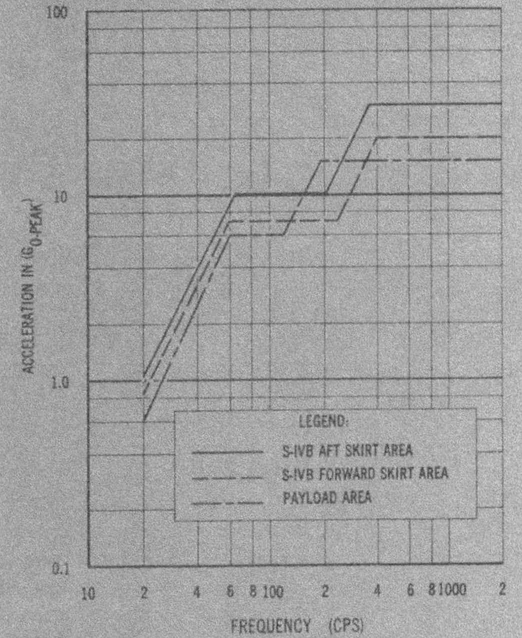

NOTES:
1. THE VIBRATION INPUT IS ASSUMED TO BE APPLIED IN EACH OF THREE MUTUALLY PERPENDICULAR DIRECTIONS.
2. THE LOGARITHMIC SWEEP RATE IS ASSUMED TO BE ONE OCTAVE PER MINUTE OVER THE FREQUENCY RANGE FROM 20 TO 2000 AND BACK TO 20 CPS.

Figure II-20
DESIGN SPECIFICATION FOR SINUSOIDAL VIBRATION

Figure II-21
S-IVB AUXILIARY
PROPULSION SYSTEM

SATURN IB

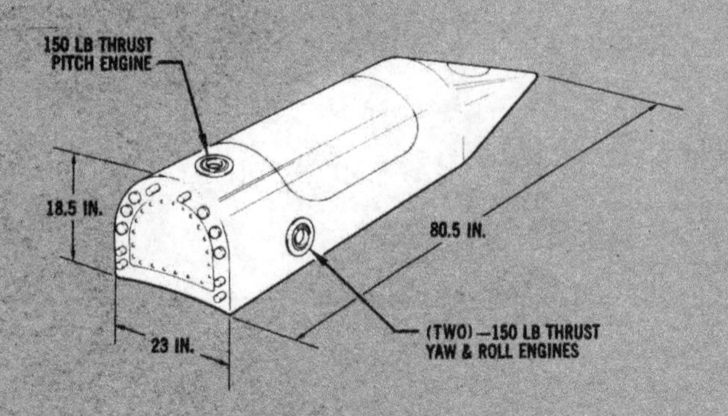

SATURN V

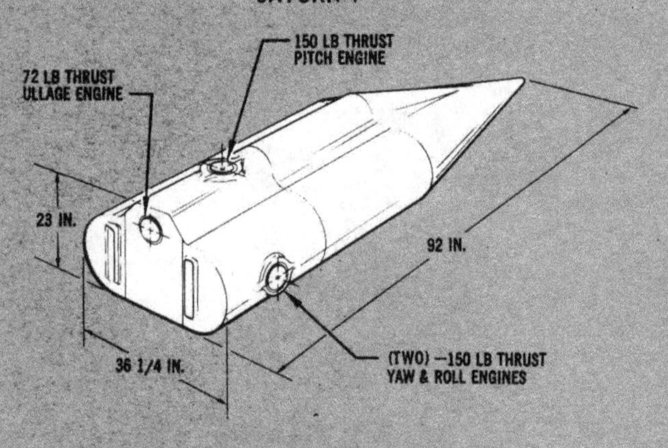

	SATURN V	SATURN IB
TOTAL MODULE DRY-WEIGHT	506 LBS	422 LBS
TOTAL WEIGHT OF LOADED MODULE	818 LBS	483 LBS
TOTAL PROPELLANT CAPACITY	312 LBS	61 LBS
MIN. TOTAL IMPULSE (0.065 SEC/PULSE)	65,000 LB-SEC	14,000 LB-SEC
NOMINAL TOTAL IMPULSE	70,000 LB-SEC	15,000 LB-SEC
MAX. TOTAL IMPULSE AVAILABLE	75,000 LB-SEC	16,200 LB-SEC

ule contains 119.4 lb of MMH (Monomethylhydrazine) fuel and 192.6 lb of N_2O_4 (nitrogen tetroxide) oxidizer. The nominal oxidizer to fuel mixture ratio is 1.65/1.

The total firing time for the engines is 7 minutes for steady state operations. Pulse operation at a pulse frequency of up to 10 pulses per second is possible. Testing has demonstrated a pulse mode capability of over 20 minutes accumulated burn time.

Many attitude control and maneuvering functions, other than those now required for the 6-1/2 hour mission, could be performed by extending the S-IVB attitude stabilization capability. An increase in the S-IVB stabilization capability would require design or operational changes to the subsystems to overcome the limitations of the present S-IVB. The major items involved in extended coast characteristics include: (a) the available mass of propellants and pressurization gases, (b) the engine life expectancy, (c) the propellant conditioning requirements to avoid freezing, (d) the attitude control dead bands, (e) the S-IVB electrical power supply, and (f) the IU electrical power supply. All of these items are closely inter-related and affect the coast time capability. If the dead band control zones were to be relaxed to $\pm 2^{\circ}$ in pitch and $\pm 10^{\circ}$ in yaw and roll, and electrical power added in the form of a fuel cell, then under certain conditions there could be sufficient propellant on-board for controlled coast times up to 30 days. Of course payload, orbit, FPR and orientation also affect coast times and must be studied for a specific mission. Consideration has been given to the above items and preliminary design concepts have confirmed that the necessary modifications can readily be made if the mission requires longer attitude-stabilized coast periods.

II-5-2. Electrical Power System

The S-IVB has four independent electrical systems with 56- and 28-volt silver-oxide primary batteries. Forward system #1 (350 ampere hours, 28 vdc) supplies power to the data acquisition system which produces low-level, high-frequency signals that must be isolated from other systems. Forward system #2 (15 ampere hours, 28 vdc) supplies power to systems which cannot tolerate switching transients or high frequency interference, such as the propellant utilization system and inverter-converter. Both batteries for the forward systems are mounted in the forward skirt.

Aft system #1 (270 ampere hours, 28 vdc) supplies power to valves, heaters and relays in the main propulsion engine, pressurization system, stage sequencer, APS modules and ullage rockets which generate switching transients, that must be isolated from other systems. Aft system #2 (70 ampere hours, 56 vdc) supplies power for an auxiliary hydraulic pump, LOX chilldown inverter and LH_2 chilldown inverter. Both batteries for aft system #1 and #2 are located in the aft skirt. Both the aft and the forward systems are wired through distribution boxes located in their respective areas.

The batteries are sized to handle the stage load requirements for 6-1/2 hours. If additional power is required for the planned 6-1/2 hours or for longer periods, additional batteries could perhaps be used for as many as 72 hours which is the wet life of the batteries. If power is required for even longer periods in orbit, other batteries or fuel cells could be used.

II-5-3. Data Acquisition System

The early Saturn V R&D vehicles have five telemetry systems; one single sideband/frequency modulation (FM) system, one pulse code modulation (PCM)/FM system, and three FM/FM systems. One channel of each FM/FM system will be used for sampled pulse amplitude modulation data. Pertinent data on these systems are shown in Table II-I.

TABLE II-I

SATURN V R & D TELEMETRY SYSTEMS

(SA-501, 502 & 503)

	T/M System	Frequency (MC/S)	Prime Channels	Prime Sampling Rate (Per sec)
1.	SS/FM	226.2	15	Continuous
2.	PCM/FM	232.9	0-100 Bi-level + Parallel Acceptance of 3 PAM Multiplexers at	120 120, 40
3.	FM/FM	246.3	15	Continuous
	PAM/FM/FM		30	120
4.	FM/FM	253.8	15	Continuous
	PAM/FM/FM		30	120
5.	FM/FM	258.5	15	Continuous
	PAM/FM/FM		30	120

Total Measurement Capability

1.	SS/FM	15 prime channels possible to sub-multiplex by 5 =	75
2.	PCM/FM	100 Bi-level channels + 30 prime channels on checkout multiplexer: 3 prime channels for frame sync & calibration; 23 prime channels possible to sub-multiplex by 10 =	234
3.	Three-5 FM/FM	15 prime channels possible to sub-multiplex by 3 =	45
	PAM/FM/FM	30 prime channels per multiplexer: 3 prime channels for frame sync & calibration; 23 prime channels possible to sub-multiplex by 10 = 234 x 3	702
			1, 056

Vehicle SA-504, to be delivered in mid-1967, and all subsequent vehicles will have only one telemetry system (PCM/FM). The capability of this operational telemetry system is:

8 channels at 120 samples/second
360 channels at 12 samples/second
190 bi-level using remote digital sub-multiplexer
558 total measurement capability

or

44 channels at 120 samples/second
190 bi-level using remote digital sub-multiplexer
234 total measurement capability

The 558 or the 234 measurement capability is based on the utilization of two multiplexers. If increased to four multiplexers, the capability becomes:

130 channels at 12 samples/second
4 channels at 120 samples/second
690 channels at 4 samples/second
12 channels at 40 samples/second
190 bi-level using remote digital sub-multiplexer
1026 total measurement capability

Figure II-22
SATURN V/S-IVB AUXILIARY
PROPULSION SYSTEM
MODULE (MOCK-UP)

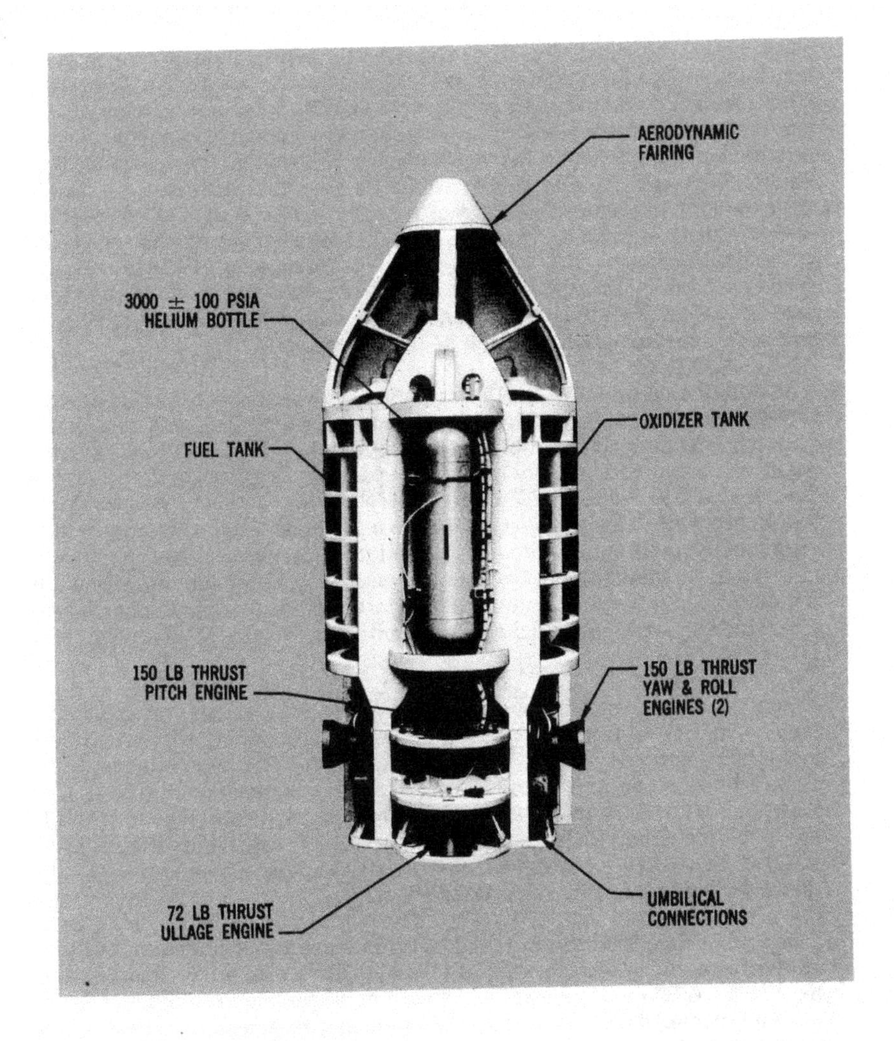

Additional combinations of sampling rates are obtainable. It is estimated that about 50 channels would be available for auxiliary payloads. The exact number of telemetry channels available can only be determined after the vehicle is selected because instrumentation varies from one vehicle to another. Once a payload application is established and scheduled, and bandwidth, accuracy, etc. are known, a determination of available channels can be made.

The operational vehicles (all vehicles after SA-503) also have provisions for mounting one complete set of modified R&D FM/FM systems in kit form. Its capability is 18 channels of continuous data, or with sub-multiplexing, 34, 66 or 82 channels depending upon the sampling rate.

Of course the payload originator may wish to furnish part or all of the Data Acquisition System associated with the experiment.

II-6. Orbital and Deep Space Tracking, Data and Control Stations

Requests for payload data must be integrated into the overall mission plan and approved by the appropriate NASA office. Orbital or space tracking and control functions required by a payload after separation from the Saturn must also be specifically approved.

II-7. Launch Support Facilities

Launch operations for the Saturn V vehicle will be conducted at Complex 39 and will utilize the mobile, or off-pad-assembly, concept. This concept, which provides for a greater flexibility and launch rate than on-pad assembly, employs four basic operations: (1) vertical assembly and checkout of the Saturn V on a mobile launcher in a controlled environment, (2) transfer of the assembled and checked-out vehicle to the launch pad on a mobile launcher, (3) automatic checkout at the launch pad, and (4) launch operations by remote control from a distant launch control center. The major units involved in this concept are the Vertical Assembly Building (VAB), Launch Control Center (LCC), Mobile Launcher (ML), Mobile Service Structure, Crawler-Transporter, Launch Pads, and High Pressure Gas Facility. Figure II-23 shows an artist's conception of the Compex 39 area. Figure II-24 shows a schematic illustration of the complex.

The Vertical Assembly Building (VAB) has two major operating areas - High Bay and Low Bay. The High Bay provides the facilities and services to assemble the complete launch vehicle in a controlled environment, and to conduct pre-launch preparations. This building is 524 feet high, 513 feet wide, and 432 feet long, and has four vehicle assembly bays and supporting facilities. Each bay is equipped with extendable platforms which are designed to permit access to the vehicle as it is assembled vertically on a mobile launcher. When the assembly of the vehicle is completed in the High Bay, pre-launch system and subsystem checks are conducted before it is moved to a launch pad.

The Low Bay area has two pairs of bays for performing continuity checks on the S-II and the S-IVB stages, engineering shops, offices, and storage space for stage pre-assembly. The S-IVB area in the Low Bay has two active stage preparation and checkout cells. This building, with its two pairs of assembly bays arranged similarly to those of the High Bay, is 118 feet high, 437 feet wide, and 256 feet long. The transfer aisle portion of the Low Bay, which connects with the High Bay transfer aisle, is 210 feet high.

The Launch Control Center, (LCC) a four-story rectangular building adjacent to the High Bay, is 76 feet high, 378 feet wide, and 181 feet long. The LCC contains offices, a cafeteria, a Complex control center, telemetry and data processing equipment for use during stage and vehicle checkouts. It also houses the firing and computer rooms which contain the control and monitoring equipment required for automatic vehicle checkout and launch.

The Mobile-Launcher (ML), upon which the Saturn V is assembled and launched, can be divided into four major elements - structure, umbilical service arms, firing accessories, and operations test and launch equipment. The structure consists of the two-story launch platform, 25 feet high, 160 feet long, and 135 feet wide, and the umbilical tower. The umbilical tower, mounted on one end of the launch platform, extends 380 feet above the deck of the structure and has eight umbilical swing arms. The arms vary in length from 35 feet to 45 feet and carry electrical, pneumatic, and propellant lines to the space vehicle. The firing accessories installed on, and considered part of the ML, include fuel fill and drain umbilicals, electrical and pneumatic umbilicals, cable masts, pneumatic-valve panels, deluge, flushing, and firefighting systems, access platforms and ladders, and a heating and ventilating system. The ML operation test and launch equipment includes a ground power system, test sets, and a computer complex.

The Mobile Service Structure is an open-frame steel truss tower designed to perform some functions at a parked position and also to be moved to the pad by the Crawler-Transporter for servicing and arming of the space vehicle. The structure is 402 feet high and is 135 feet by 132 feet at the base.

A Crawler-Transporter is used to position the Mobile Launcher (ML) in the Vertical Assembly Building (VAB), to move the ML space vehicle

Figure II-23
SATURN V ON PAD 39-KSC

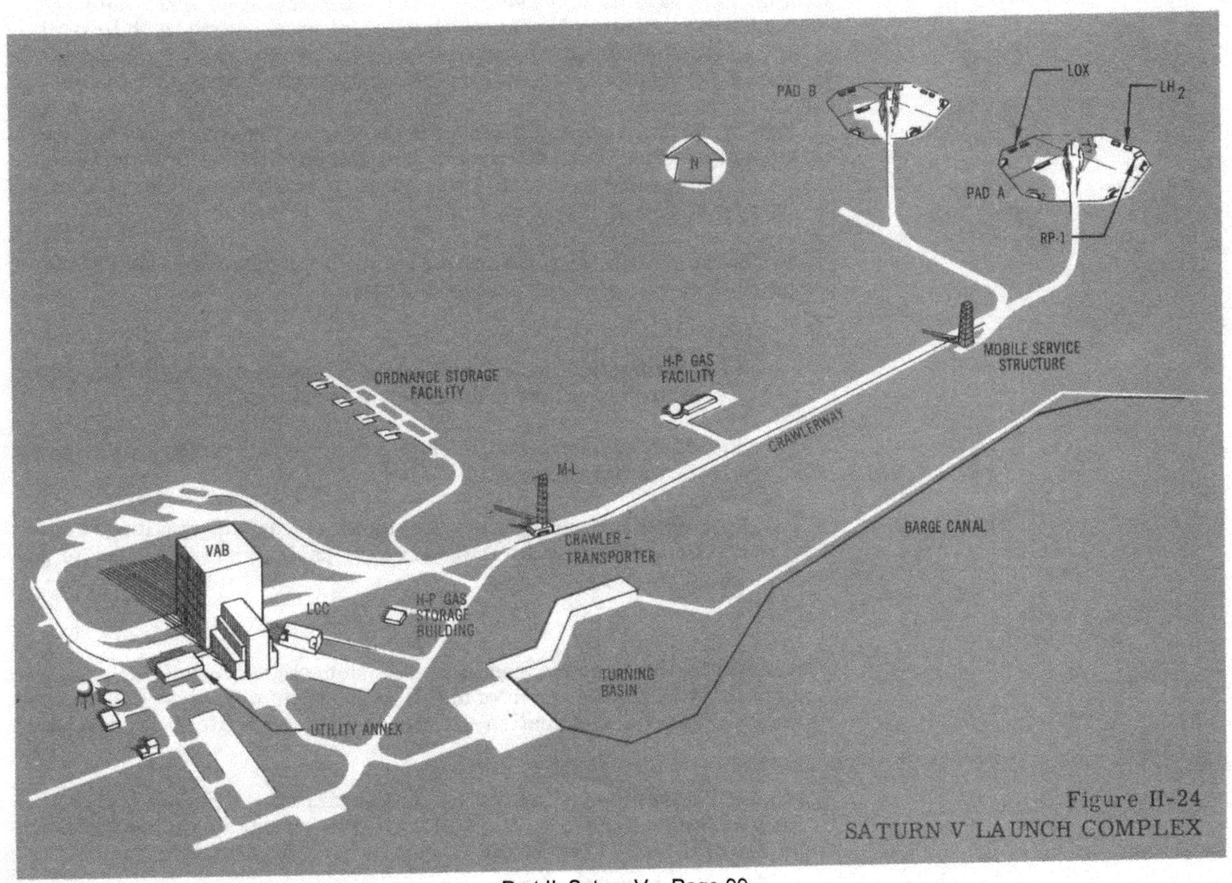

Figure II-24
SATURN V LAUNCH COMPLEX

configurations from the VAB to the launch pad, and to move the Mobile Service Structure from its parked position to the launch pad. It incorporates a large platform and four tractor units as a self-contained vehicle. This equipment, 131 feet long, 114 feet wide, and weighing 5.5 million pounds, is powered by diesel generators developing a total of over 7000 kilowatts for motivation, leveling, and the steering system. The leveling system keeps the ML and the Saturn V Vehicle within one-sixth of a degree of true vertical while negotiating a curve of up to five hundred feet in radius and a 5 per cent grade. The Crawler-Transporter, which can be steered from either end, has a normal loaded speed of one mph maximum and an unloaded speed of two mph maximum.

Each launch pad of Complex 39 is in the shape of an eight-sided polygon, with a distance across (perpendicular to crawlerway) of 3000 feet. The overall hardstand area of 390 feet by 325 feet has a center portion elevated 42 feet to allow sufficient clearance for positioning a two-way flame deflector beneath the ML after it is anchored for launch.

Installed at the launch pad at varying distances from the hardstand area are the propellant storage and transfer facilities for LO_2, LH_2, and RP-1 (900,000 gallons for LO_2, 850,000 gallons for LH_2 and 250,000 gallons for RP-1).

Those accommodations which are required to support the launch of an auxiliary payload must be arranged for, or provided by the experiment sponsor, unless available equipment can be used without conflict. It is imperative that the auxiliary payload planner evaluate, at an early date, the ground support equipment and range support required to launch his experiment. Small payloads can often be accommodated within the existing equipment and facilities. Special calibration, checkout, alignment, and handling equipments which are peculiar to the auxiliary payload will require early planning and arrangements. Small payloads can often be handled with relative ease with mobile trailers which are brought in by the payload agency. In the event that building space is required, special arrangements must be made in advance.

One of the earliest actions involved in staging the launch of any payload is the preparation of the Range Data and Support Requirements Document. If separate auxiliary and primary payload documents are required, they must be coordinated.

In the Block House, specific control of the experiment countdown may be handled in one of several ways, namely:

(a) Automatic countdown control of auxiliary payload functions with manual override from payload console.

(b) Manual checkout and control of auxiliary payload functions from payload console.

(c) Automatic integrated countdown of auxiliary payload functions with launch vehicle or primary payload with no separate payload console.

Obviously, the specific requirements and objectives of the auxiliary payload will dictate which of the above modes of operation will be used. Again, early coordination with the launch vehicle is important.

Space can be provided near the Launch Complex for payload originators to park instrumented trailers of their own for remote radio-line checkout if available checkout facilities are not adequate.

II-8. Data Reduction and Evaluation

Prime payloads (mounted above the S-IVB stage and the I.U.) generally will have self-contained telemetry systems; whereas, auxiliary payloads could use available S-IVB or I.U. Telemetry systems. The following information is needed to insure proper use of vehicle telemetry:

(a) Type of measurement (pressure, temperature, signal, vibration, strain, etc.)

(b) Range and accuracy of measurement needed

(c) Type of monitoring (continuous, sampling, real time)

(d) Type of presentation (punch tape, magnetic tape or strip chart)

The interconnect between the test article and the stage will be made by standard flight proven methods. Where appropriate, transducers and signal conditioning devices will be existing flight qualified items.

The worldwide tracking network is utilized for orbital and deep space operations. The tracking network is composed of ground stations around the world which have been established for various space missions. These stations provide the capability of tracking, data acquisition, and communications for space programs. For manned Earth orbital and Lunar operations, the Manned Spaceflight Network or the Space Tracking and Data Acquisition Network would normally be used. For planetary missions, the Deep Space Instrumentation Facility is used. The use of these networks must be arranged in advance with NASA. Procedures exist which will permit data to be recorded in the form of tapes and strip charts from NASA, USAF, and other organizations.

A limited amount of data may be received prior to lift-off via stage umbilicals and could be recorded on strip charts, sequence recorders or magnetic tape as required. All data telemetered from the vehicle is received and recorded during pre-launch, launch, and orbit. Vehicle data may aid reduction and interpretation of payload data. Quick look and in some cases, real time data can be provided by the ETR Facilities.

Douglas can give experimenters partial or complete reduction and evaluation of data in a final report. Douglas has data reduction facilities at the Douglas Huntington Beach Data Laboratory which can reduce data to the following forms:

(a) Analog strip charts

(b) Tabulated digital readouts in engineering units

(c) Plots of digital data in engineering units

(d) Analog oscillograph plots

(e) Digital magnetic tapes in engineering units

The Huntington Beach Computer Facility is equipped with IBM 7094 computers for lengthy, iterative, computation processes.

II-9. Checklist for Auxiliary Payload/Vehicle Interface Requirements

The first important step in planning an auxiliary or prime payload for Saturn V is to document information on the physical and operating characteristics of the payload along with the required launch vehicle accommodations and ground support. With such information, it will be possible for you to discuss the various aspects of Saturn V flight accommodations with appropriate planning agencies. A typical check list is given below of the items of information required to properly consider and define Launch Vehicle accommodations for your payload. Information on the experiment submittal process and associated vehicle data can also be obtained from cognizant NASA Agencies.

A. General Information Required

Experiment title

Proposal Originator

Purpose and application of experiment

Relationship to Apollo or other national goals

Description of experimental procedures

Present status of experimental equipment

Scope of budget or available funding

B. Experiment Mission Requirements

Orbital Altitude; circular, elliptical (apogee-perigee)

Synchronous orbit/Hohmann transfer

(1) circularization by S-IVB

(2) circularization by payload

Suborbital or orbital flight durations, minimum, maximum

Desired launch azimuth

Desired launch inclination

Desired date of launch (year)

Astronauts' time required, pre-flight, inflight, post-flight

C. Experimental Equipment Capability or Requirements

Envelope description or volume requirements

Weight

Environmental Limitations or Capability

(1) temperature

(2) acoustics

(3) vibration

(4) shock

(5) acceleration

(6) humidity and free moisture

(7) atmosphere and pressure

(8) sand and dust

(9) meteoroids

(10) fungus

(11) salt spray

(12) ozone

(13) hazardous gases

(14) particle radiation

(15) electromagnetic radiation

(16) electromagnetic compatibility

(17) explosion proofing

(18) sterilization requirements

(19) special environmental control

Electrical Power Loads; voltage, current, duration, AC-DC

(1) steady state

(2) intermittent

(3) peak

(4) desired interface locations

Vehicle gas requirements; flowrates, pressures, temperatures

(1) helium

(2) nitrogen

(3) oxygen

(4) hydrogen

(5) others

Jettison Requirements

Special Attitude Control Requirements

(1) stabilization-control precision

(2) angular acceleration and velocity in pitch, yaw and roll

Schedule Information

Range Safety Requirements

. Instrumentation Requirements

Type and Numbers: Pressure, temperature, signal, vibration, strain, special

Range and Accuracies

Type of Monitoring

(1) continuous

(2) sampling

(3) real time

Duration or Time period of monitoring

Interface

(1) transducer part of experimental package

(2) transducer part of stage contractor responsibility

(3) location

E. Final Data

Raw data desired

Reduced data desired

Evaluated data desired

Final data package, reports, tapes, graphs, etc.

F. Shroud Design for Prime Payloads

Configuration A, B, C, D, or special (Figures II-11, 12, 13 and 14)

G. Suggested Mounting Location

H. Ground Support Equipment (Location and Type)

Electrical checkout

Pneumatic

Mechanical

Handling

Servicers

I. Tracking, Data Acquisition and Command

J. Facilities

Facilities needed by payload originator at Douglas Space Systems Center or Kennedy Space Center.

K. Special

Any special requirements that affect the integration of the payload with the launch vehicle.

If you desire help in integrating your experiment with the Saturn V vehicle, please forward your request to the address shown in the foreword of this guide.

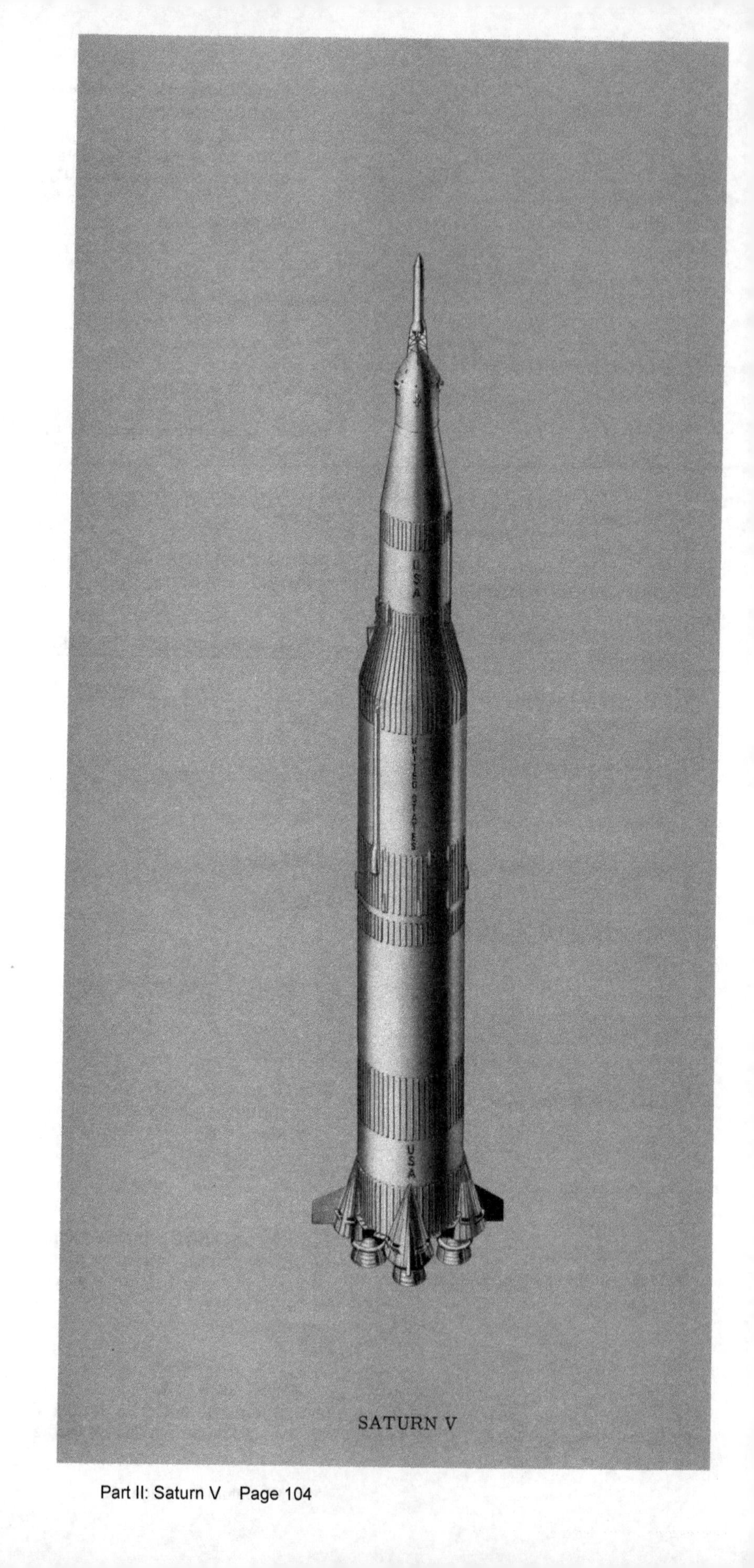

SATURN V

The Saturn V Vehicle is a three-stage configuration designed primarily for accomplishing the Apollo Lunar Landing Mission. This vehicle, shown in Figure III-1, consists of (a) the S-IC first stage, built by The Boeing Company, (b) the S-II second stage, built by North American Aviation, (c) the S-IVB third stage, built by the Douglas Missile and Space Systems Division and (d) an Instrument Unit (IU) built by International Business Machine Corporation. The IU is mounted above the S-IVB and houses the guidance, control and non-stage oriented flight instrumentation. The payload shown above the IU is the Apollo system for the Lunar Landing mission. The payload is made up of a Command Module (CM), a launch escape system, a service module, a Lunar Excursion Module (LEM) and a LEM Adapter which houses the ascent and descent stages of the LEM.

The Saturn V Vehicles represent the largest launch vehicle under development in the United States. Its relationship to the earlier Saturn configurations is shown in Figure III-2. The first Saturn V launch vehicle will be flown in the later-half of 1966. The three stage standard Saturn V, with payload, weighs approximately 3200 tons at liftoff, can place up to 261,000 pounds of payload into a 100 nautical mile circular earth orbit and can accelerate 98,000 pounds to escape velocity. The first-stage engines generate a total of 7,500,000 pounds of thrust at sea level.

The Saturn V Vehicles stand alone in their payload class and have practically unlimited applications to both manned and unmanned earth orbital, Lunar or interplanetary missions. Although a single large payload may be the primary purpose for launching the Saturn V Vehicle, many auxiliary scientific or engineering payloads can also be carried into space economically aboard the Saturn upper (S-IVB) stage of the Saturn.

III

SATURN V CONFIGURATION

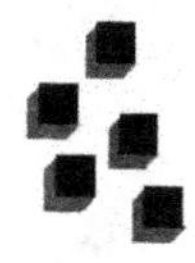

III-1. LOR/Apollo Configuration

III-1-1. First Stage (S-IC)

The S-IC stage, shown in Figures III-3 and III-4, the first stage of the Saturn V launch vehicle, is manufactured by The Boeing Company at the Michoud Plant near New Orleans, Louisiana. The first four development stages were built by MSFC at Huntsville, Alabama. The stage uses five Rocketdyne F-1 engines, each of which produces a nominal thrust of 1.5 million pounds and uses a mixture of Liquid Oxygen (LOX) and RP-1 (special kerosene fuel) as a propellant. The five engines burn for 150 seconds and lift the vehicle to an altitude of approximately 30 nautical miles before burnout occurs. Four of the engines are gimbal-mounted on a 364 inch diameter circle and are hydraulically gimballed to provide thrust vector control in response to steering commands from the guidance system located in the Instrument Unit.

The stage utilizes separate propellant tanks that are all welded assemblies of cylindrical ring segments with dome-shaped end bulkheads. Each tank has slosh baffles over the full depth of the liquid. The LOX tank is pressurized by gaseous oxygen while the RP-1 tank uses stored helium.

Eight 80,000 pound-thrust retro-rockets provide separation of the S-IC from the S-II Stage. These solid propellant motors are mounted in pairs under each engine fairing.

III-1-2. Second Stage (S-II)

The second stage of the Saturn V Vehicle is the S-II (Figures III-5 and III-6) which is being developed by North American Aviation's Space and Information System Division at Downey, California. The stage

Figure III-1
OPERATIONAL SATURN V CONFIGURATIONS

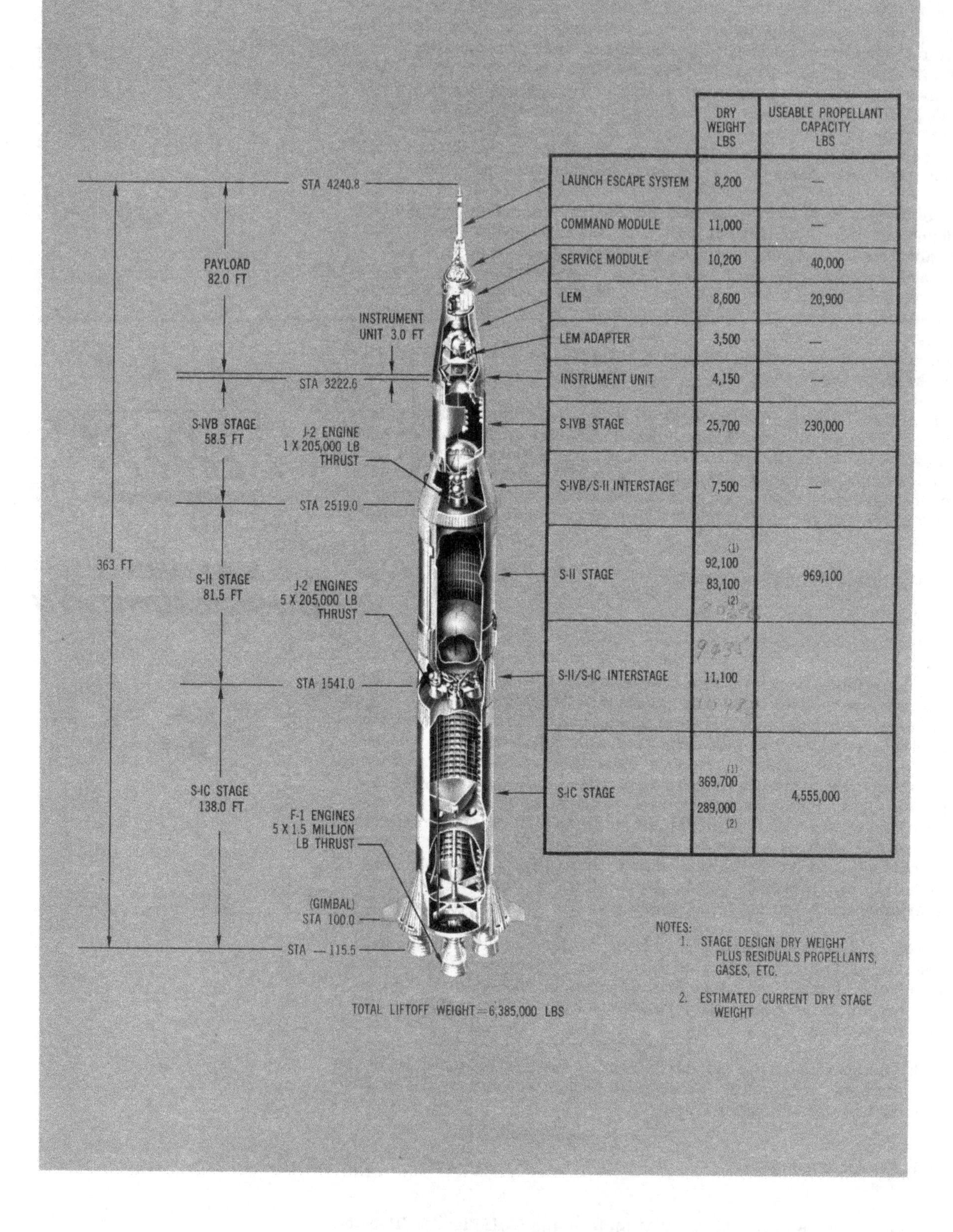

	DRY WEIGHT LBS	USEABLE PROPELLANT CAPACITY LBS
LAUNCH ESCAPE SYSTEM	8,200	—
COMMAND MODULE	11,000	—
SERVICE MODULE	10,200	40,000
LEM	8,600	20,900
LEM ADAPTER	3,500	—
INSTRUMENT UNIT	4,150	—
S-IVB STAGE	25,700	230,000
S-IVB/S-II INTERSTAGE	7,500	—
S-II STAGE	(1) 92,100 83,100 (2)	969,100
S-II/S-IC INTERSTAGE	11,100	
S-IC STAGE	(1) 369,700 289,000 (2)	4,555,000

NOTES:
1. STAGE DESIGN DRY WEIGHT PLUS RESIDUALS PROPELLANTS, GASES, ETC.
2. ESTIMATED CURRENT DRY STAGE WEIGHT

uses five Rocketdyne J-2 engines, each rated at a nominal 205,000 pound-thrust, and burns a mixture of liquid oxygen and liquid hydrogen The propellants are contained in a cylindrical tank with domes at each end and an insulated common bulkhead to separate the upper LH_2 tank from the LOX tank. Each tank contains slosh baffles to minimize propellant slosh. Eight solid propellant 22,900 pound-thrust rocket motors burning for 3.74 seconds are used to ullage the propellants for engine start.

The five S-II engines burn for about 375 seconds and boost the vehicle to an altitude of approximately 100 nautical miles. Four of the engines are gimbal-mounted and are hydraulically gimballed to provide thrust vector control in response to steering commands from the guidance system located in the Instrument Unit.

III-1-3. Third Stage (S-IVB)

The third stage of the Saturn V is the S-IVB (Figures III-7 and III-8) which is being developed by the Douglas Missile and Space Systems Division at Huntington Beach, California.

The S-IVB has a single 205,000 pound thrust Rocketdyne J-2 engine that burns liquid oxygen (LOX) and liquid hydrogen (LH_2). The Saturn V/S-IVB as presently designed has a 4-1/2 hour orbital plus a 2 hour translunar coast capability. The tankage contains 230,000 pounds of usable propellant at a LOX to LH_2 mass ratio of 5 to 1.

The thrust is transmitted to the stage through a skin and stringer structure shaped in the form of a truncated cone that attaches tangentially to the aft liquid oxygen dome. The hydrogen tank is internally insulated with reinforced polyurethane foam and contains a series of high pressure spheres, storing gaseous helium, for liquid oxygen tank pressurization. Adapter structures, referred to as the forward and aft skirt and the aft interstage, provide the necessary interfaces for mating with the payload and the lower stages. The tank structure features a waffle-like pattern on the hydrogen tank sidewall to act as a semi-monocoque load bearing member. A double walled composite structure with an insulating fiberglass honeycomb core forms the

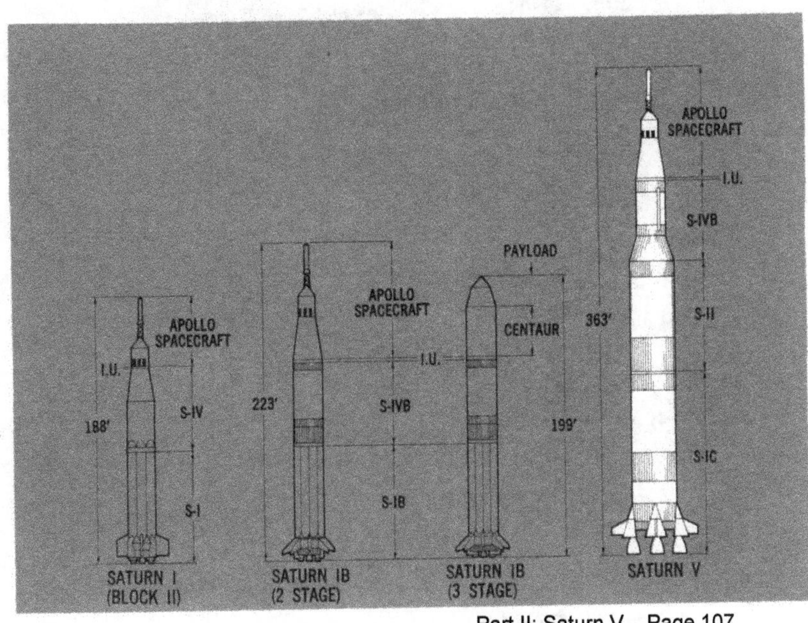

Figure III-2
SATURN LAUNCH VEHICLES

Figure III-3
SATURN V/S-IC STAGE

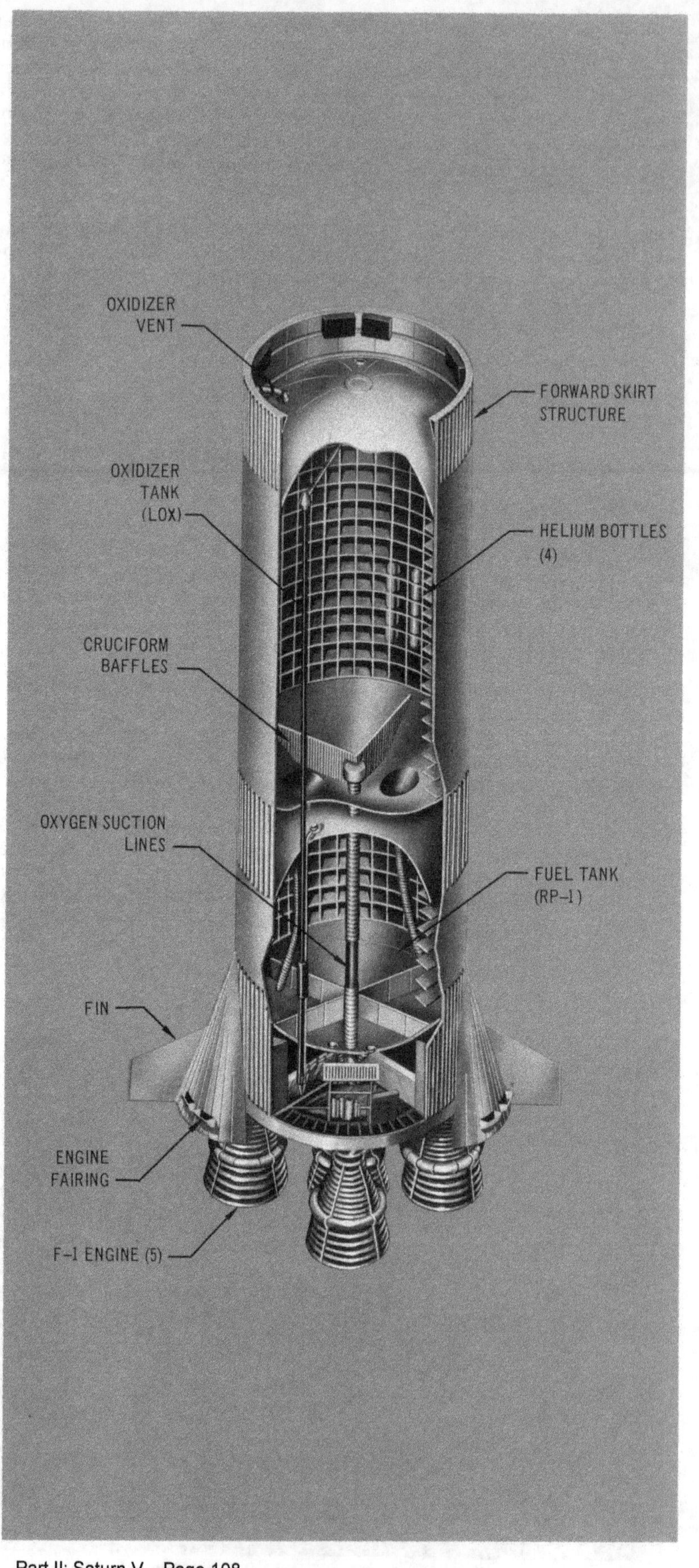

Figure III-4
S-IC STAGE INBOARD PROFILE

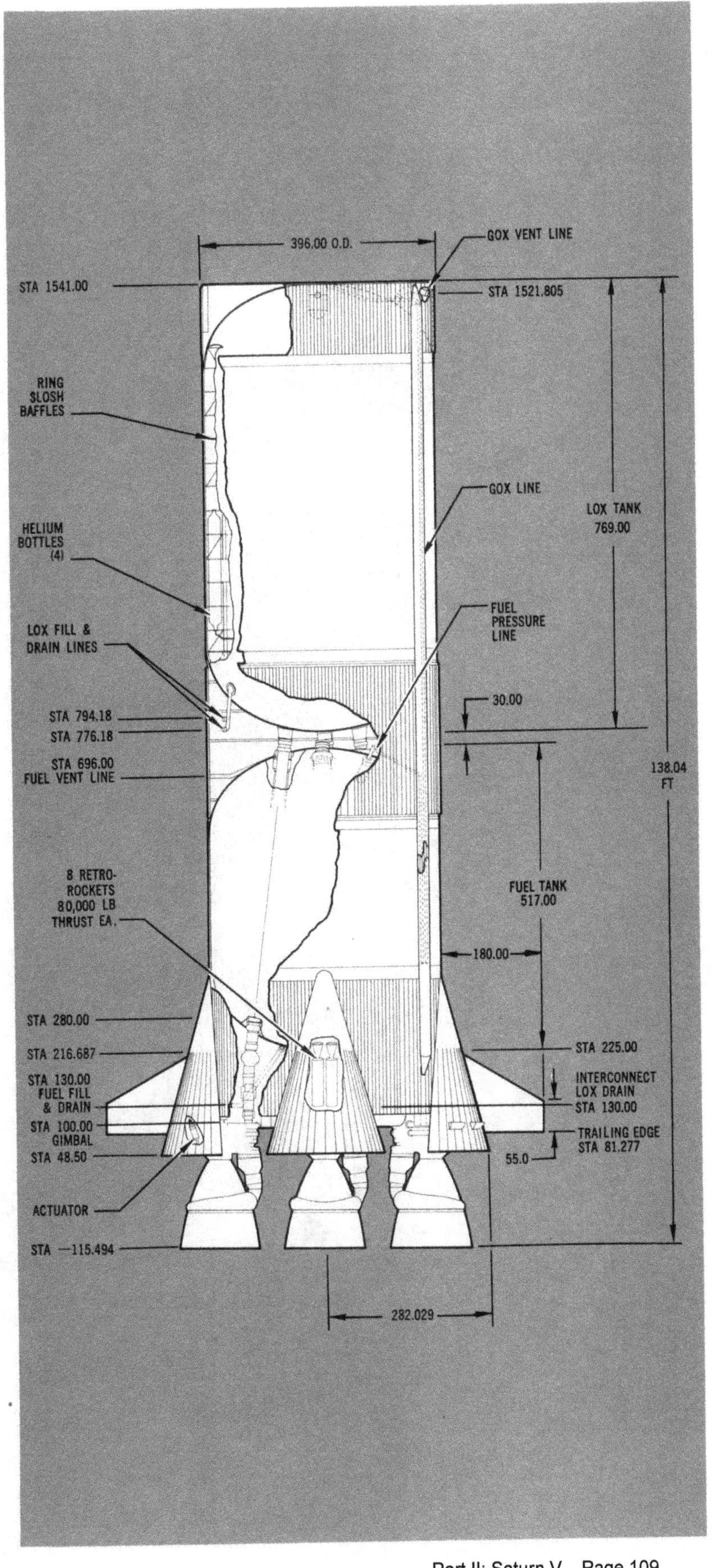

Figure III-5
SATURN V/S-II STAGE

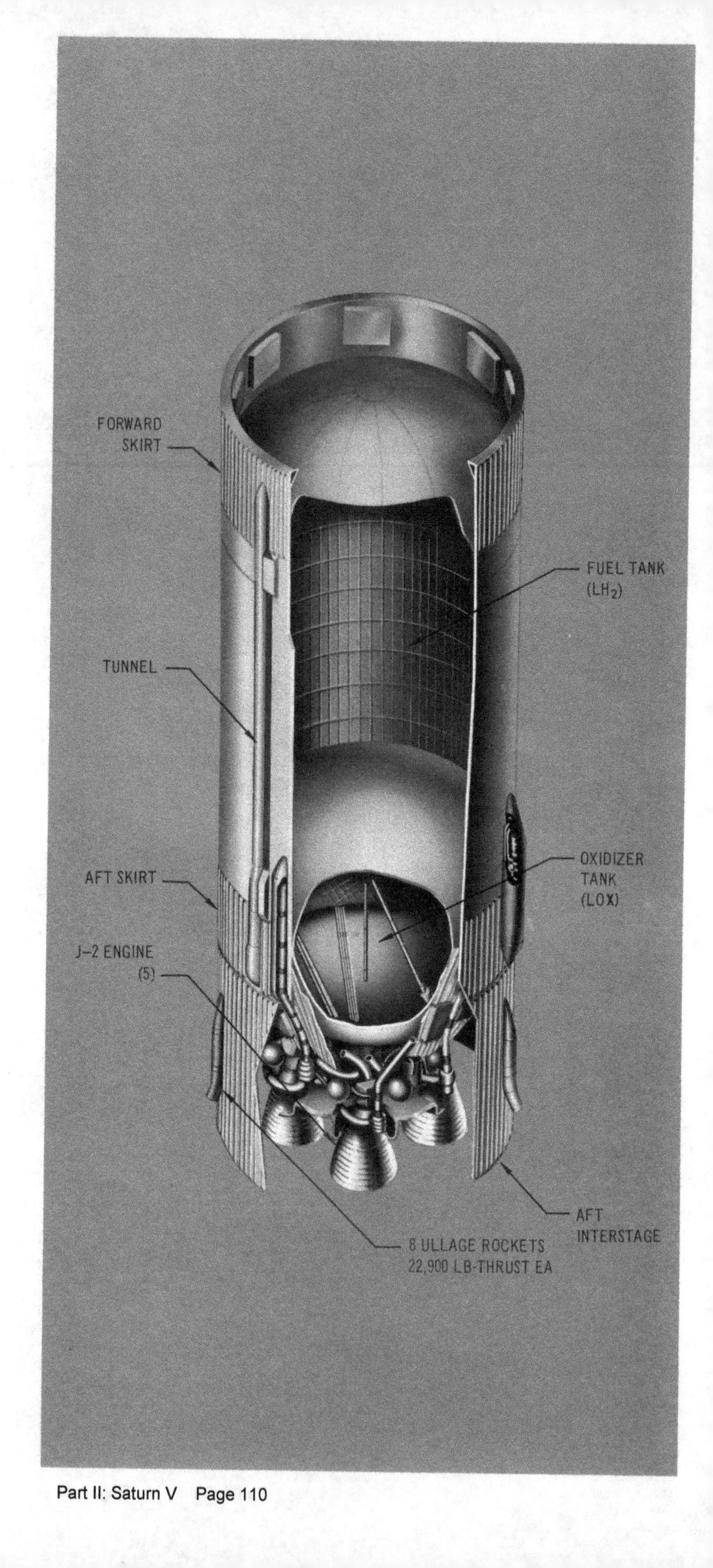

Figure III-6
SATURN V/S-II
STAGE PROFILE

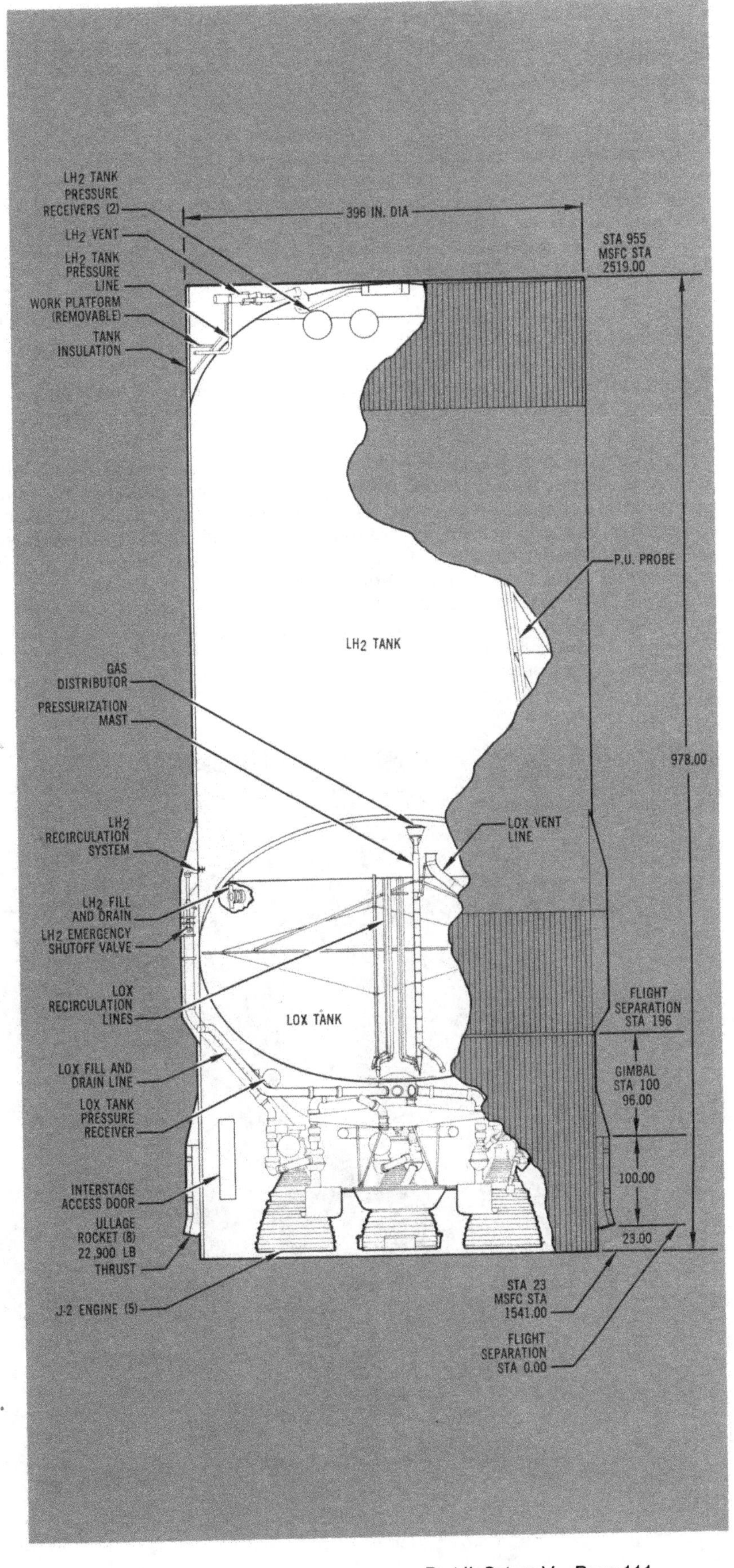

common bulkhead which separates the hydrogen and oxygen tanks. The propellant tanks have spherical end domes. Skirt and interstage structures are composed of conventional skin, external stringers and internal frames.

Pitch and yaw attitude are controlled during powered flight by gimballing the main engine. Roll control is provided by 150-pound thrust engines located in the Auxiliary Propulsion System (APS) modules. Three axis (roll, pitch, and yaw) attitude control during orbital and translunar coast or unpowered flight is provided entirely by the APS. The signals for vehicle attitude control originate in the guidance and control system located in the instrument unit.

The APS modules are located on the aft skirt assembly of the S-IVB, 180° apart from each other and utilize nitrogen tetroxide (N_2O_4) and monomethylhydrazine (MMH) as the propellant. Each Saturn V/S-IVB module has two 150-pound thrust roll/yaw engines, one 150-pound thrust pitch control engine and one 72-pound thrust ullage engine.

The separation of the S-IVB from the S-II is initiated by an explosive charge which parts the aft skirt from the aft interstage. Two 3400-pound thrust solid propellant ullage rockets mounted on the S-IVB are then ignited and burn for approximately 4 seconds to settle the propellant in the tanks by maintaining a positive acceleration. Four 35,000-pound thrust solid retro-rockets located on the aft interstage are fired simultaneously for 1.5 seconds to decelerate the first stage. The J-2 is ignited 1.6 seconds after separation signal and is at full thrust within 5 seconds of the ignition signal. The J-2 burn-into-orbit is about 152 seconds in duration. The two 72-pound ullage APS motors

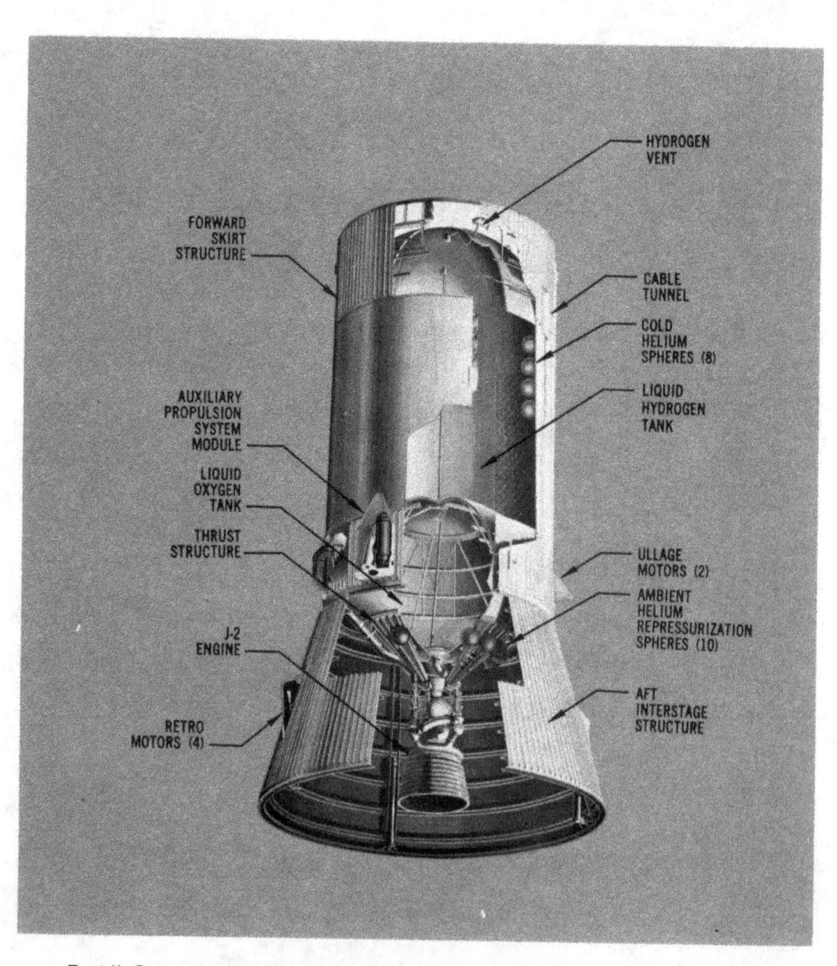

Figure III-7
SATURN V/S-IVB STAGE

SATURN V/S-IVB STAGE
INBOARD PROFILE

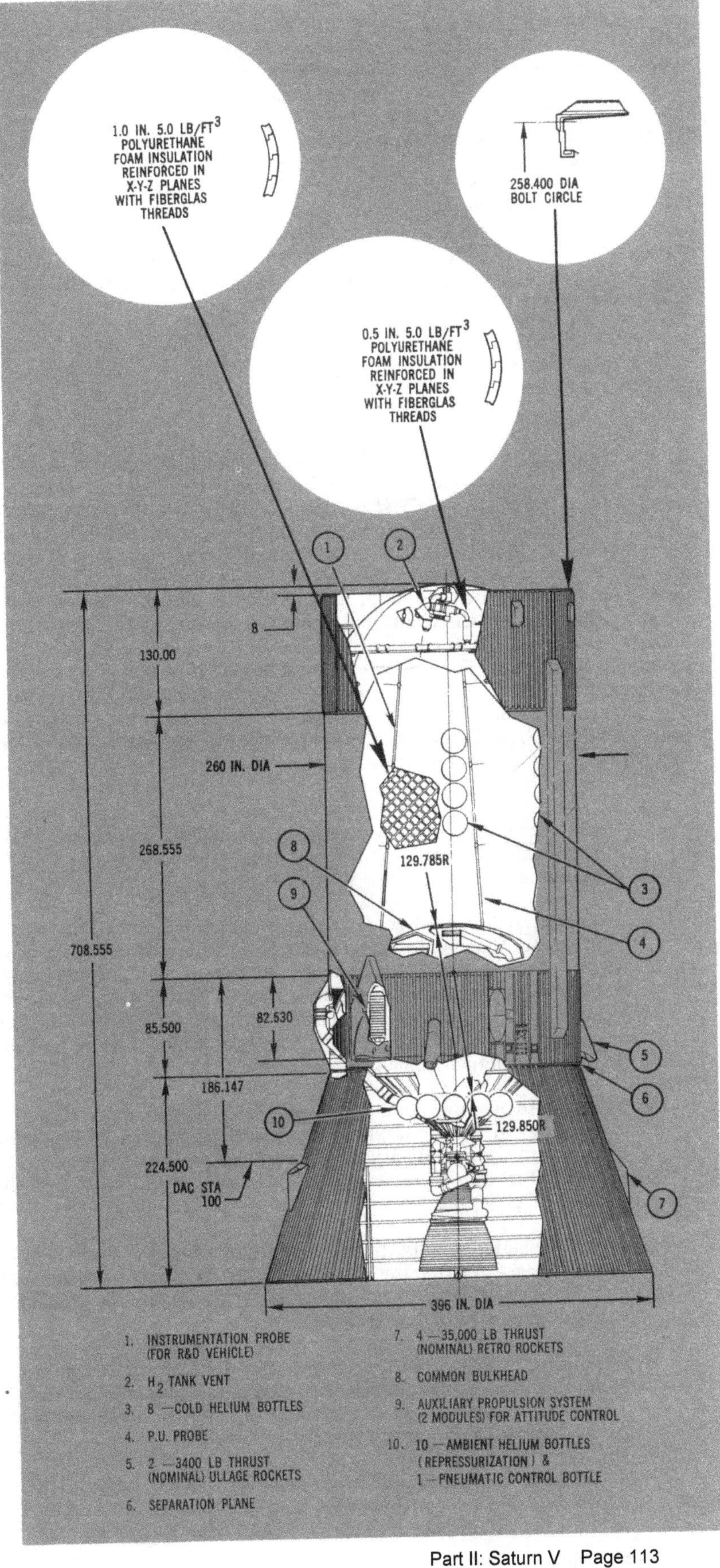

are burned for about 90 seconds during the J-2 shutdown to maintain propellant control. The ullage engines are fired again for 327 seconds during J-2 engine chilldown prior to the second J-2 start. Ten ambient-temperature helium spheres provide gas for repressurizing the LH_2 and LOX tanks for the restart. A weight saving modification to the stage is in progress which will utilize cold helium stored in bottles within the hydrogen tank and an oxygen-hydrogen burner for the repressurization.

The J-2 engine burns a second time for about 339 seconds to put the payload and S-IVB stage into a translunar trajectory. Engine shutdown is triggered by the guidance system when orbit insertion velocity is achieved, then three axis attitude stabilization is maintained as described above.

III-1-4. Instrument Unit (I. U.)

The instrument unit, fabricated by International Business Machines Corporation, is a 260-in. diameter by 36-in. high cylindrical section located forward of the S-IVB (Figure III-9). The I.U. is designed for a 6-1/2 hr. orbital and translunar coast capability but could be modified for longer durations. The electrical and environmental control systems are the limiting systems. This 3990 lb unit, which is the "nerve center" of the launch vehicle, contains the guidance system, the control systems and the flight instrumentation systems for the launch vehicle. Access to the inside of the S-IVB forward skirt area is provided through an I.U. door. Electrical switch selectors provide the communications link between the I.U. computer and each stage. The computer controls the mode and sequence of functions in all stages. The I.U. consists of six major subsystems as listed below in (a) through (f).

(a) The structural system or the aluminum cylindrical body of the unit which carries the payload and supports the various systems.

(b) The environmental control system provides electronic equipment cooling during ground operations and throughout flight. The coolant is a 60%-40% methanol-water mixture which circulates through a series of cold plates. In flight, the absorbed heat is removed through a heat exchanger that vents boiled-off water to space. For ground operation, the system rejects heat to a thermo-conditioning servicer.

(c) The guidance and control systems provide guidance and control sensing, guidance steering computations, and control system signal shaping and summing. The shaped control signals are fed to the appropriate actuating devices on the S-IC, S-II and the S-IVB stages.

(d) The measuring and telemetry system transmits signals from the vehicle or experiment transducers during ground check-out and flight to ground command stations by various frequency bands and modulating techniques.

(e) Radio Frequency (RF) systems maintain contact between the vehicle and ground stations for tracking and command purposes. They consist of Azusa and C-band transponders, and an S-band command receiver and transmitter.

(f) A separate electrical system generates and distributes 28 vdc power required for operation of all of the above systems. Some of this power may be available to experimenters.

III-2. Man Rating, Reliability and Quality Control

One prime objective in the design of the Saturn S-V vehicle is to provide a safe, reliable vehicle able to carry a variety of manned and unmanned spacecraft into many different orbits and space trajectories. The approach used to achieve this objective is to impose stringent man-rating, reliability, and quality control procedures throughout the design, production and checkout of the vehicle.

III-2-1. Man Rating

The S-V vehicle provides an Emergency Detection System (EDS) for automatic failure warning and also automatic mission abort capability should the crew have insufficient time to react to the failure. When sufficient time exists, the EDS provides the crew with data displays enabling them to decide whether or not to abort. It is designed to minimize the possibility of automatically aborting because of a false signal. The EDS and abort procedures are closely integrated with range safety procedures to ensure that the crew can escape safely. In addition, a Malfunction Detection System (MDS) is being developed for use on the S-V vehicle.

The principle of the EDS is being expanded into an MDS so that more parameters are monitored, and the crew is provided with additional data displays. The mission go/no-go decision capability of the crew is increased. The safety features of the MDS include those of the EDS, in addition to a greater number of data displays providing the capability of verifying the out-of-tolerance condition of a parameter. This latter capability is added protection against the false abort mode.

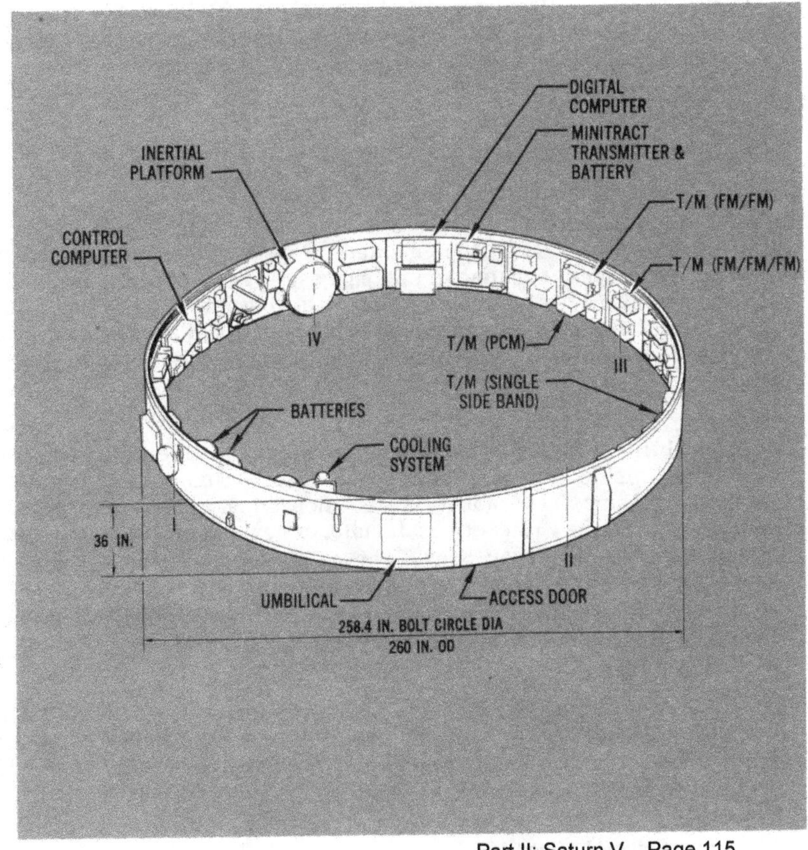

Figure III-9
INSTRUMENT UNIT

III-2-2. Reliability

High reliability of components, subsystems, and systems are a basic design parameter of the Saturn Program. The program for attaining and maintaining high reliability consists of the following elements:

(a) Failure effect analysis of the design.

(b) A thorough and complete test program.

(c) Imposition of stringent reliability and quality control procedures.

The failure effect analysis consists of a detailed technical analysis of the design of the system to identify all the possible significant failure modes, categorizing the effects of each failure mode, and elimination of failure modes.

Thorough component, subsystem and system tests are conducted in the laboratory, at the Static Test Facility and during prelaunch checkout. Post-flight data evaluation of vehicle systems serve as a tool to assess reliability for future missions. Carefully planned procedures and controls used in these tests and data correlations establish a measure of reliability and determine the level of confidence in the measure. All these factors help meet the Saturn V reliability goal of 0.90. The reliability goal of the S-IVB stage is 0.95 at a 90% confidence level. Experimental payloads will be given the same attention to ensure the same high probability of success.

Stringent quality control standards in manufacture, fabrication, and testing ensure that reliability will not be degraded by human error or by manufacturing techniques. NASA documents of the NPC 200 series, (Quality Program Provisions) and NPC 250-1, (Reliability Provisions for Space System Contractors), contain the reliability requirements and quality control standards which guide payload planners.

III-2-3. Quality Control

Strict quality control standards including a comprehensive failure reporting system are implemented to assure non-degradation of reliability during manufacturing phases. Failure data is used to update reliability estimates and improve design. In addition, the failure data is automated and used throughout the test and checkout program to provide rapid, comprehensive reporting of the failure history of the vehicle.

A traceability program has also been implemented for the vehicle on all items which could cause an aborted mission during final countdown or a flight failure. Traceability is also imposed on those items utilizing new processes, new or exotic materials, or new and unique applications of old materials and processes. These items are serialized, lot coded or date coded and evidence that all inspections and test operations have been performed is retrievable. Thus, traceability requirements provide evidence of proper configuration of critical components.

The high reliability of the booster vehicle should be matched by high payload reliability. Therefore to provide assurance of the experiment's success, the reliability requirements imposed on the experiment and/or the payload must be at least equal to that of the S-IVB Stage.

IV-1. LOR/Apollo Mission Profile

The Saturn V three stage launch vehicle is being developed to provide a booster system for the Apollo Lunar Orbital Rendezvous (LOR) and Landing mission. That is, the Saturn V is designed to fulfill the mission velocity requirements of the Apollo mission up to injection of the Apollo spacecraft and Lunar Excursion Module (LEM) into a 72-hour lunar transfer trajectory. This mission requires two stages and a portion of the propellant from the third stage to achieve a one-hundred nautical mile circular orbit. A coast of up to three revolutions in this orbit (4-1/2 hours) is allowed for vehicle checkout and to determine the precise achieved orbit for third-stage engine restart operations. At the proper time, the third (S-IVB) stage re-ignites and boosts the spacecraft into the lunar transfer trajectory. During the translunar flight following S-IVB main engine shutdown, the spacecraft/launch vehicle separation occurs, followed by spacecraft transposition, docking, and midcourse orbit corrections. A sequence of events summary for the LOR/Apollo mission is illustrated in Figure IV-1. Upon reaching the vicinity of the moon, the spacecraft is injected into a parking orbit about the moon and the LEM separates, and descends to the the lunar surface.

IV-2. General Three-Stage Mission Profile

SATURN V PERFORMANCE

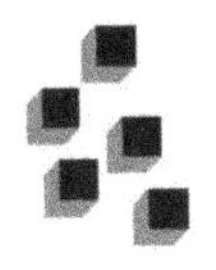

The Saturn V three-stage launch vehicle is capable of placing heavy payloads in various circular, elliptical, or hyperbolic orbits. The vehicle capabilities and flight profiles for such missions are described in the following paragraphs:

For three-stage missions, the launch sequence is initiated with the ignition of the five first-stage engines. The vehicle rises vertically to clear the umbilical tower and rolls to align the pitch plane with the desired launch azimuth. Initiation of a pitch program then starts the vehicle down range. The pre-programmed pitch attitude history is designed to follow a ballistic trajectory (zero angle of attack) under no-wind conditions. At 154.6 seconds after liftoff, the center engine is shut down and four seconds later the remaining four control engines are shut down. Three and eight-tenths seconds are then required to separate the empty S-IC stage and reach 90% thrust on the S-II stage. After approximately thirty seconds, to ensure vehicle attitude stabilization, the forward section of the two-piece S-IC/S-II interstage is jettisoned. At approximately this same time, the launch escape system will be jettisoned on manned flights.

The S-II stage, which uses a programmed propellant mixture ratio to optimize the engine thrust/specific impulse history, reaches propellant depletion and is separated from the S-IVB stage approximately 536 seconds after liftoff. The second and third stage attitude history is determined by an iterative guidance scheme based on the calculus-of-variations which minimizes the propellant burned in reaching the desired burnout velocity and position. A command cutoff occurs upon injection of the S-IVB/Payload into the desired orbit. If this orbit is a parking orbit (intermediate to a terminal point), the S-IVB will be reignited at some time later and will propel the payload to the desired final conditions of velocity and position.

Table IV-1 gives the weight breakdown of the Saturn V stages for the three-stage launch vehicle. The S-IVB propellant figure shown represents the total tank capacity. The actual amount consumed may be less and is dependent upon the specific mission.

TABLE IV-I

SATURN V VEHICLE WEIGHT SUMMARY

	Weights Lb.
S-IC at Separation	381, 645
S-IC Stage/Residuals	(369, 700)
S-IC/S-II Aft Interstage (1)	(1, 330)
S-IC/S-II Separation/Start Losses	(10, 615)
S-IC Propellant	4, 555, 003
S-IC/S-II Forward Interstage (1)	9, 770
S-II at Separation	100, 664
S-II Stage/Residuals	(92, 139)
S-II/S-IVB Interstage	(7, 468)
S-II/S-IVB Separation/Start Losses	(1, 057)
S-II Propellant	969, 078
S-IVB at Separation	28, 549
S-IVB Dry Stage (2)	(25, 708)
S-IVB Residuals	(2, 841)
S-IVB Total Usable Propellant Capability	230, 000
Flight Performance Reserves (3)	(2, 907)
S-IVB Weight Loss in Parking Orbit (4)	3, 495
Instrument Unit	4, 150
	6, 282, 354 (5)

Note: (1) Two plane separation - 1,330 pounds separates with S-IC. 9,770 pounds is carried with S-II for 30 seconds before jettison.

(2) Includes 204 pounds of jettisoned weight (ullage cases, etc.).

(3) Typical - 0.75 per cent of vehicle characteristic velocity, included in S-IVB propellant usable capacity.

(4) Typical for 4-1/2 hour coast.

(5) Total weights do not include payload, payload adapter, shroud, or launch escape system. Weights based on projected data for Vehicle SA-504.

Figure IV-2 presents the circular orbit capabilities for the three-stage vehicle for direct ascent missions to various orbit altitudes and inclinations when launched from the Eastern Test Range (ETR). Orbital payload capability, via Hohmann transfer, is shown in Figure IV-3 for the case of a due east launch. These data are based on the assumptions that the launch site is in the plane of the desired orbit and no trajectory plane-changing ("dog-leg") maneuvers are per-

Figure IV-1
SATURN V NOMINAL
LOR/APOLLO MISSION
SEQUENCE OF EVENTS

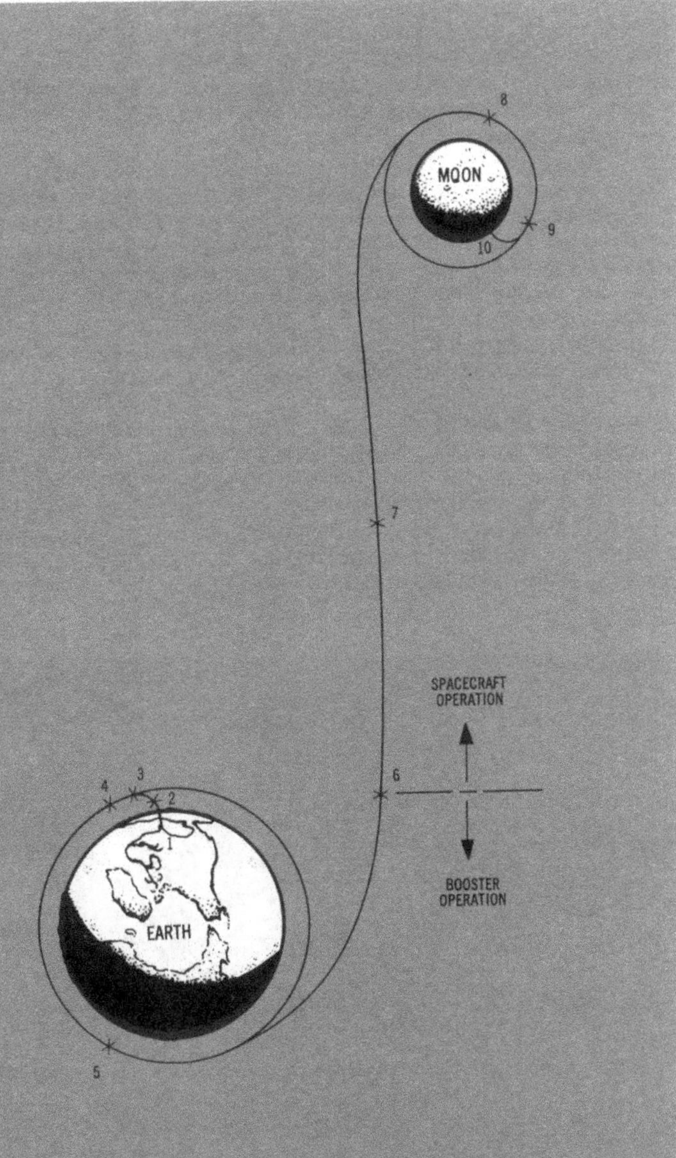

EVENTS:

1. LAUNCH FROM E.T.R.
2. S-IC BURNOUT AND JETTISON S-II IGNITION
3. S-II BURNOUT AND JETTISON S-IVB FIRST IGNITION
4. INJECTION INTO EARTH PARKING ORBIT S-IVB FIRST SHUT DOWN
5. ORBITAL LAUNCH S-IVB SECOND IGNITION
6. LUNAR TRANSFER ORBIT INJECTION S-IVB SECOND SHUTDOWN, APOLLO TRANSPOSITION, DOCKING, AND S-IVB JETTISON
7. MID-COURSE CORRECTIONS
8. INJECTION INTO LUNAR PARKING ORBIT
9. DE-ORBIT LUNAR EXCURSION MODULE
10. LUNAR EXCURSION MODULE TOUCHES DOWN

formed. The approximate sector of allowable launch azimuths without requiring a "dog-leg" maneuver is between 40 to 140 degrees (launch orbit inclinations greater than approximately 55 degrees, or launch azimuths less than 40 degrees or greater than 140 degrees, would require special range safety waivers or a "dog-leg" ascent with a resultant decrease in payload).

Figure IV-4 shows the payload capabilities to a synchronous orbit (24-hour period) via Hohmann transfer as a function of orbit inclination for selected launch azimuths. The required plane rotation is accomplished coincident with circularization at apogee. Capabilities for a 60 degree inclined synchronous orbit are shown in Figure IV-5 as a function of launch azimuth. These missions require a third start capability which the S-IVB presently does not have. Weight penalties associated with a third start have been included.

Figure IV-6 shows the payload capability of the three-stage vehicle to various elliptical orbits for a due east launch. The interplanetary mission capability of the Saturn V Vehicle is shown in Figure IV-7. These data are based on a due east launch to a 100 nautical mile parking orbit and an orbit launch by the S-IVB. Some representative trajectory parameter histories for a high energy mission are also shown in Figure IV-8 through IV-14.

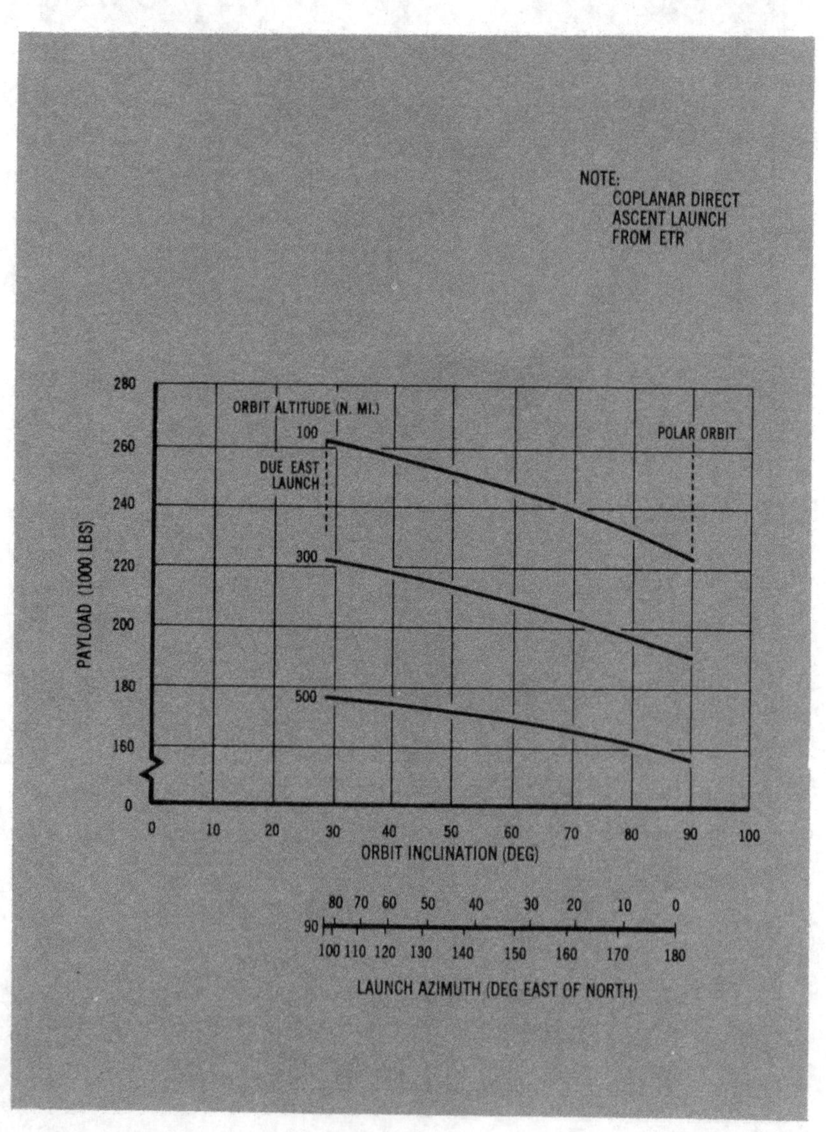

Figure IV-2
SATURN V CIRCULAR
ORBIT CAPABILITY

Figure IV-3
SATURN V HOHMANN TRANSFER PAYLOAD CAPABILITY

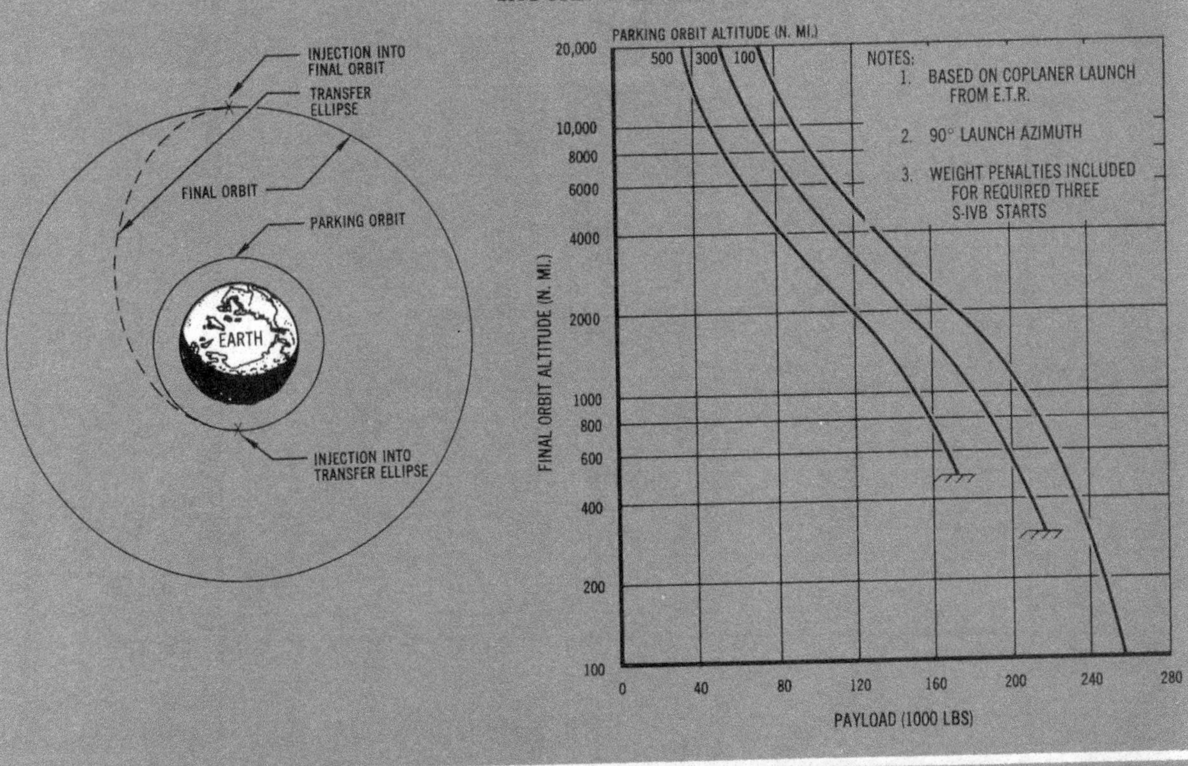

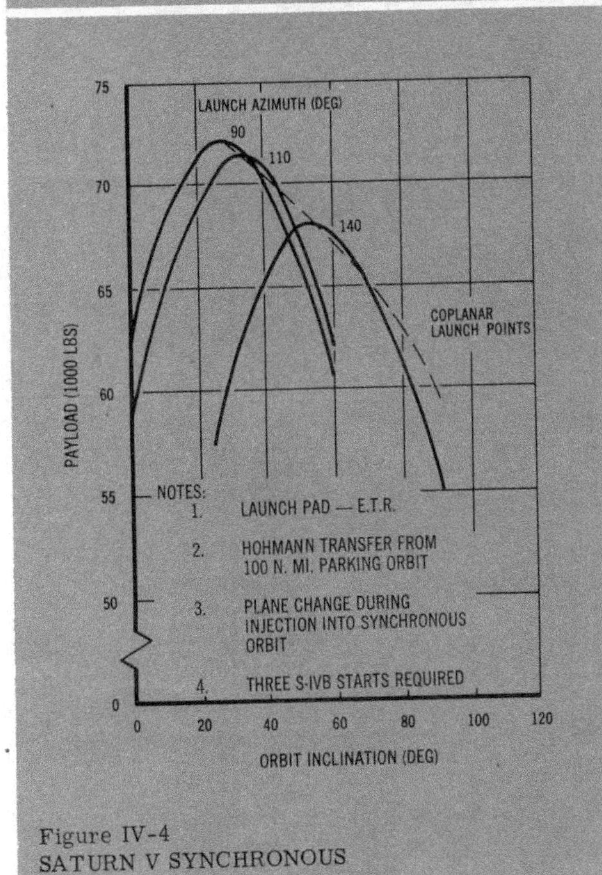

Figure IV-4
SATURN V SYNCHRONOUS ORBIT CAPABILITY

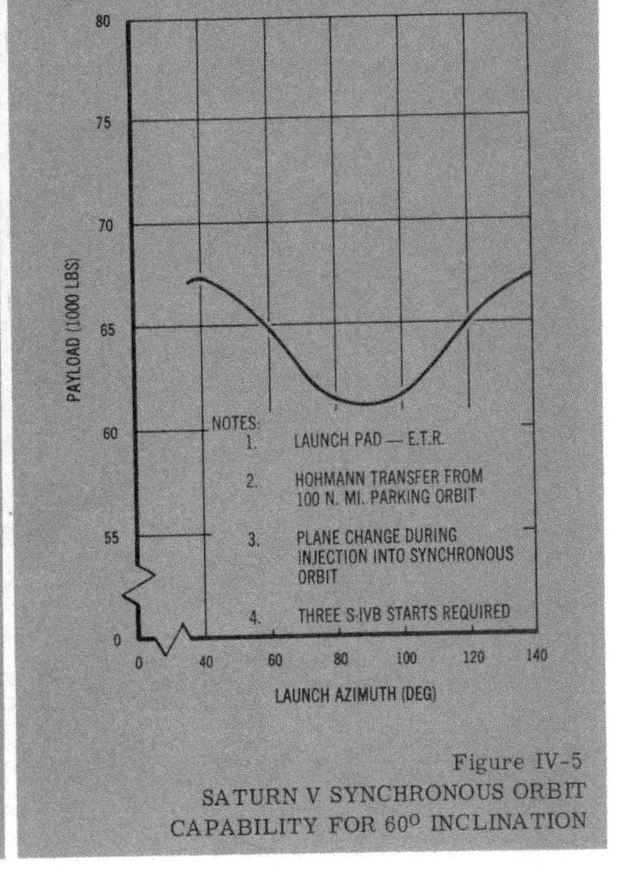

Figure IV-5
SATURN V SYNCHRONOUS ORBIT CAPABILITY FOR 60° INCLINATION

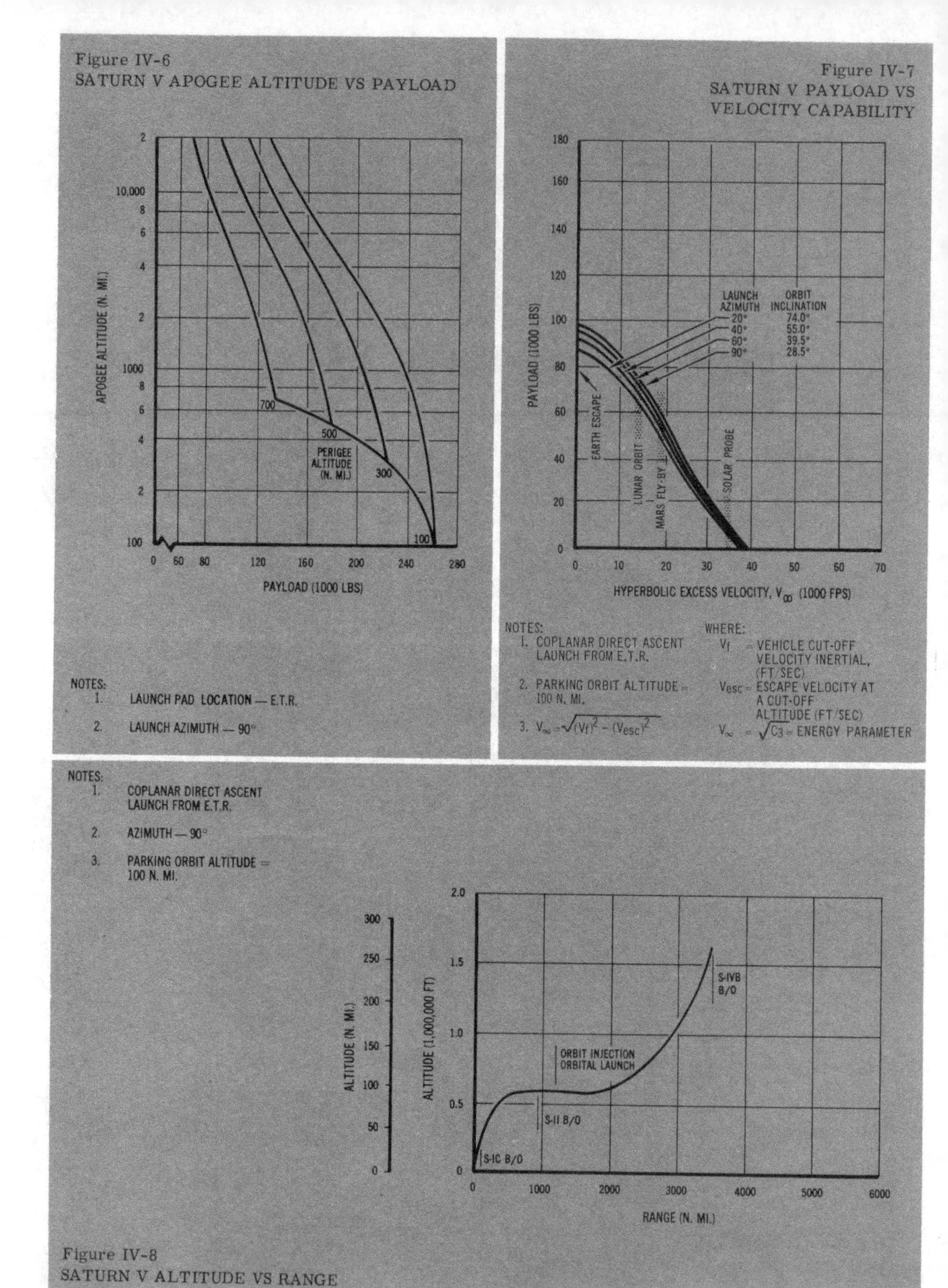
Figure IV-6
SATURN V APOGEE ALTITUDE VS PAYLOAD
APOGEE ALTITUDE (N. MI.)
PAYLOAD (1000 LBS)
PERIGEE ALTITUDE (N. MI.)
700
500
300
100
NOTES:
1. LAUNCH PAD LOCATION — E.T.R.
2. LAUNCH AZIMUTH — 90°
Figure IV-7
SATURN V PAYLOAD VS VELOCITY CAPABILITY
PAYLOAD (1000 LBS)
HYPERBOLIC EXCESS VELOCITY, V∞ (1000 FPS)
LAUNCH AZIMUTH
ORBIT INCLINATION
20° 74.0°
40° 55.0°
60° 39.5°
90° 28.5°
EARTH ESCAPE
LUNAR ORBIT
MARS FLY-BY
SOLAR PROBE
NOTES:
1. COPLANAR DIRECT ASCENT LAUNCH FROM E.T.R.
2. PARKING ORBIT ALTITUDE = 100 N. MI.
3. $V_\infty = \sqrt{(V_f)^2 - (V_{esc})^2}$
WHERE:
V_f = VEHICLE CUT-OFF VELOCITY INERTIAL, (FT/SEC)
V_{esc} = ESCAPE VELOCITY AT A CUT-OFF ALTITUDE (FT/SEC)
$V_\infty = \sqrt{C_3}$ = ENERGY PARAMETER
NOTES:
1. COPLANAR DIRECT ASCENT LAUNCH FROM E.T.R.
2. AZIMUTH — 90°
3. PARKING ORBIT ALTITUDE = 100 N. MI.
ALTITUDE (N. MI.)
ALTITUDE (1,000,000 FT)
RANGE (N. MI.)
S-IC B/O
S-II B/O
ORBIT INJECTION ORBITAL LAUNCH
S-IVB B/O
Figure IV-8
SATURN V ALTITUDE VS RANGE

Figure IV-9
SATURN V DYNAMIC PRESSURE
VS FLIGHT TIME

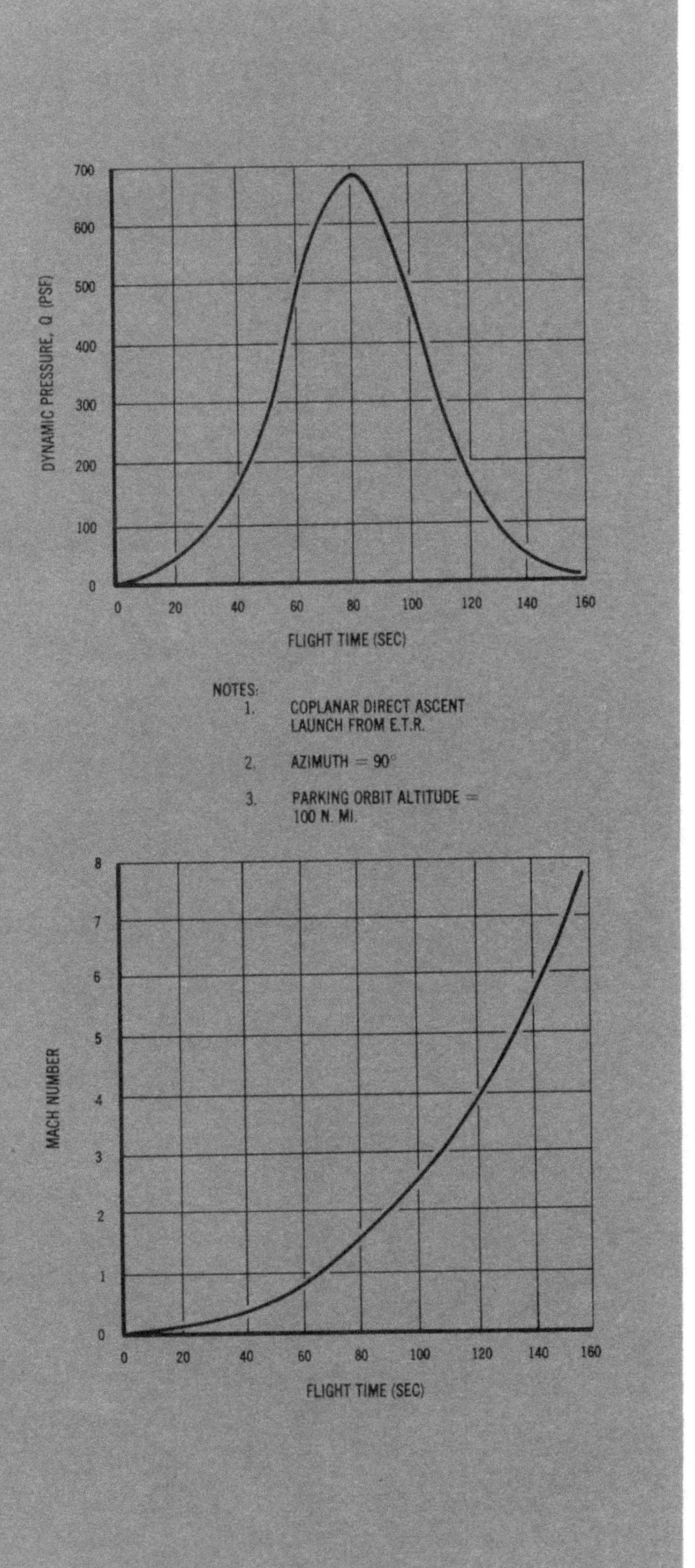

Figure IV-10
SATURN V MACH NUMBER
VS FLIGHT TIME

Figure IV-11
SATURN V INERTIAL FLIGHT PATH ANGLE VS FLIGHT TIME

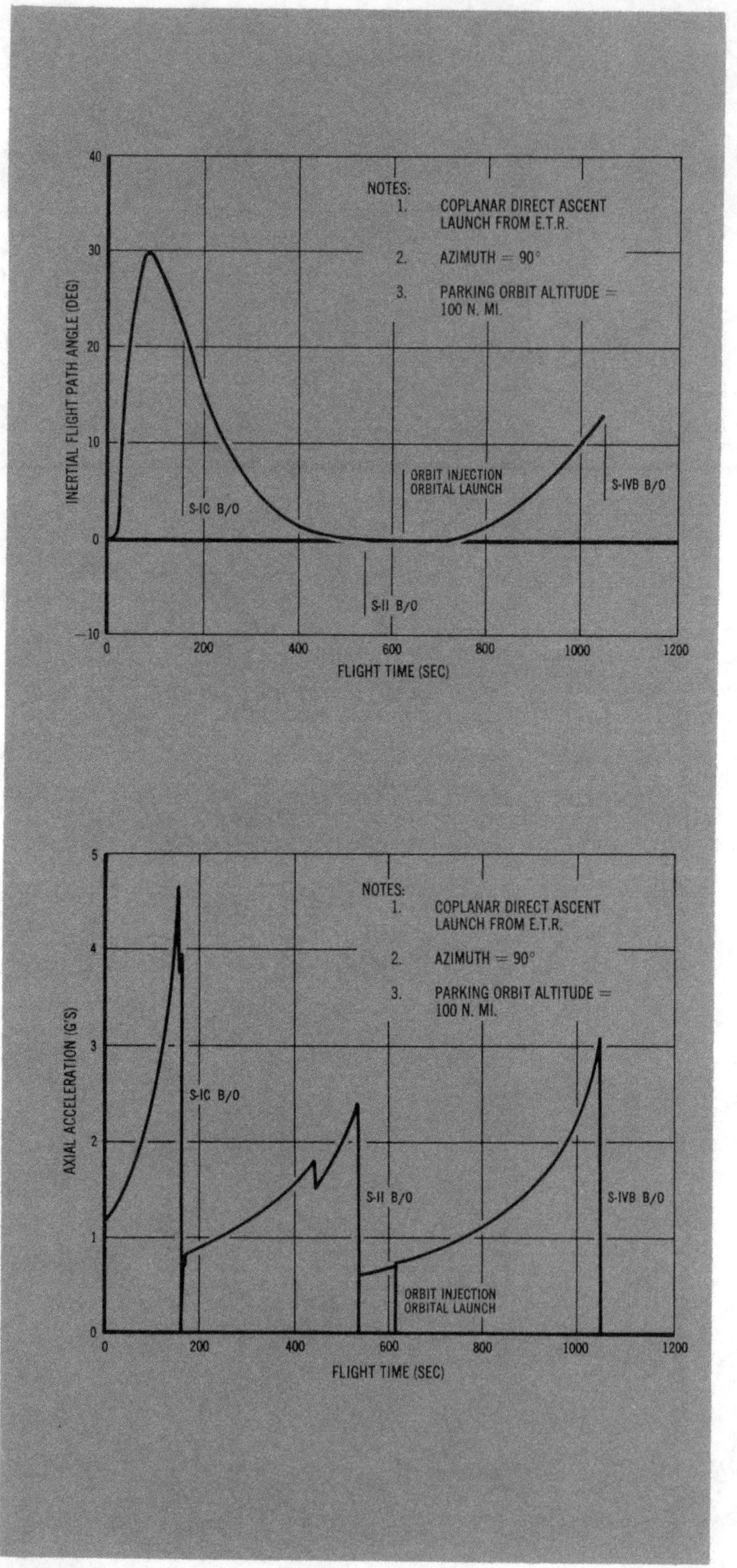

Figure IV-12
SATURN V AXIAL ACCELERATION VS FLIGHT TIME

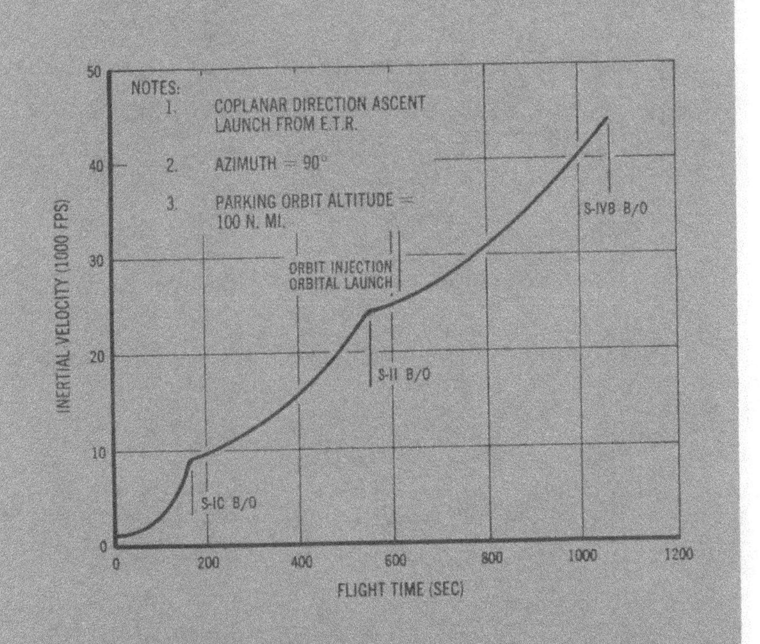

Figure IV-13
SATURN V INERTIAL VELOCITY VS FLIGHT TIME

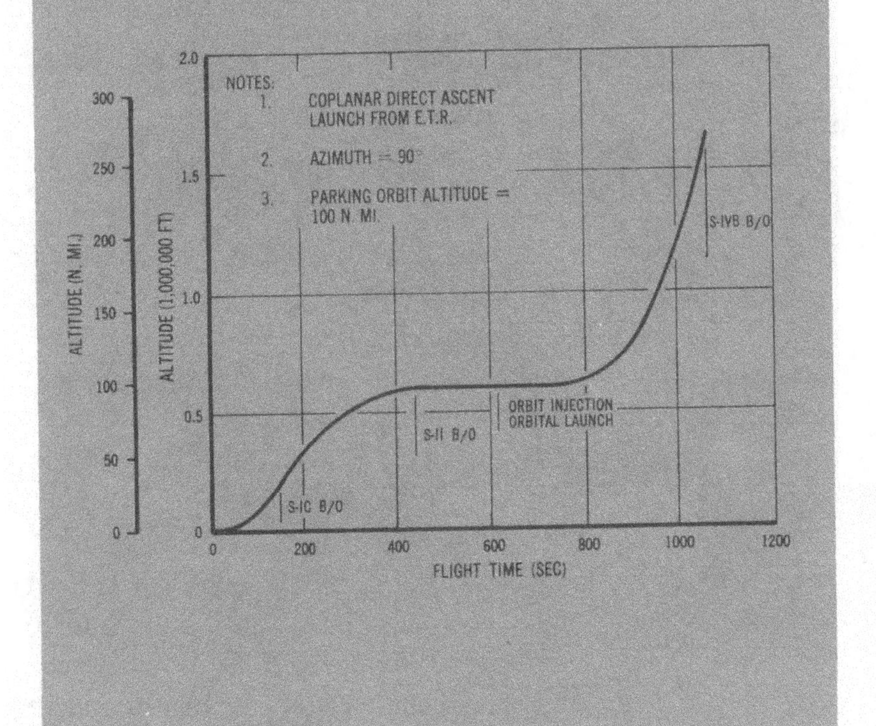

Figure IV-14
SATURN V ALTITUDE VS FLIGHT TIME

USA

V-1. Growth Configurations of the Saturn V

The three-stage configuration and performance capabilities described in this document are based on present estimates of a standard operational Saturn V Vehicle. Means of improving the basic vehicle to achieve higher performance are continually being studied. The various performance improvement techniques include such items as (a) adding engines to the S-IC and S-II, (b) increasing the thrust of the F-1 and J-2 engines, (c) increasing the propellant capacities of each of the three stages, (d) using improved engines on the S-II and S-IVB with higher thrust and performance and (e) using solid motor strap-ons on the S-IC.

Other ways to improve the payload delivery capabilities and the versatility of the Saturn V will undoubtedly be studied for some time. The present operational vehicle can launch a payload weighing over 261,000 pounds to a 100 n. mi. circular orbit and 98,000 pounds to escape velocity. Improved performance studies indicate that payloads of 450,000 pounds to a 100 n. mi. circular orbit and 170,000 pounds to escape velocity are possible. The results of Saturn V improvement studies have shown that the Saturn V class of vehicles has substantial growth potential for future missions.

V-2. High Energy Mission Vehicles

SATURN V GROWTH POTENTIAL

The exploration of the solar system and the space beyond presents an exciting challenge to the scientific and engineering community. High spacecraft velocities or energies are required to explore the solar system or to perform deep space missions. The present Saturn V, even with the performance improvement techniques described above, is not capable of performing certain high energy missions; therefore, a high energy upper stage must be added. The Saturn V, with an additional upper stage, can offer the highest payload potential of any vehicle under development for these high energy missions. The Saturn V/Centaur is an example of a vehicle representing near term availability. The Centaur would serve as a high energy fourth-stage on the Saturn V as depicted in Figure V-1. The four-stage mission profile is similar to the three-stage mission up to the one-hundred nautical mile parking orbit. However, the shroud enclosing the Centaur is jettisoned at an altitude of approximately 340,000 feet, 214.5 seconds after liftoff. The S-IVB re-ignites in orbit, expends its propellant and is jettisoned. The Centaur stage is ignited and propels the payload to the desired final conditions.

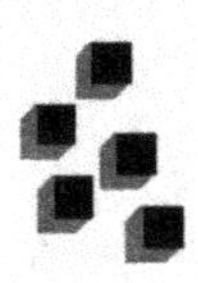

Table V-I gives the weight breakdown of the Saturn V stages of the four-stage launch vehicle. The S-IVB and Centaur propellant data shown represent tank capacities. The actual amount needed would depend on the specific mission. In determining the performance of the four-stage vehicle, an S-IVB propellant loading of 230,000 pounds was used. This is not optimum for all missions and some small gain in performance may be achieved through optimization (within the tank capacity limits) of this parameter for a specific mission. The Saturn V/Centaur Vehicle has the capability to perform the types of missions as listed in Table V-II and illustrated in Figure V-2. Representative performance curves and trajectory parameter histories for a high energy mission are shown in Figures V-3 through V-8 for the four-stage Saturn V/Centaur Vehicle launched due east from the Eastern Test Range through a 100 nautical mile parking orbit.

Figure V-1
HIGH ENERGY MISSION
SATURN V CONFIGURATION

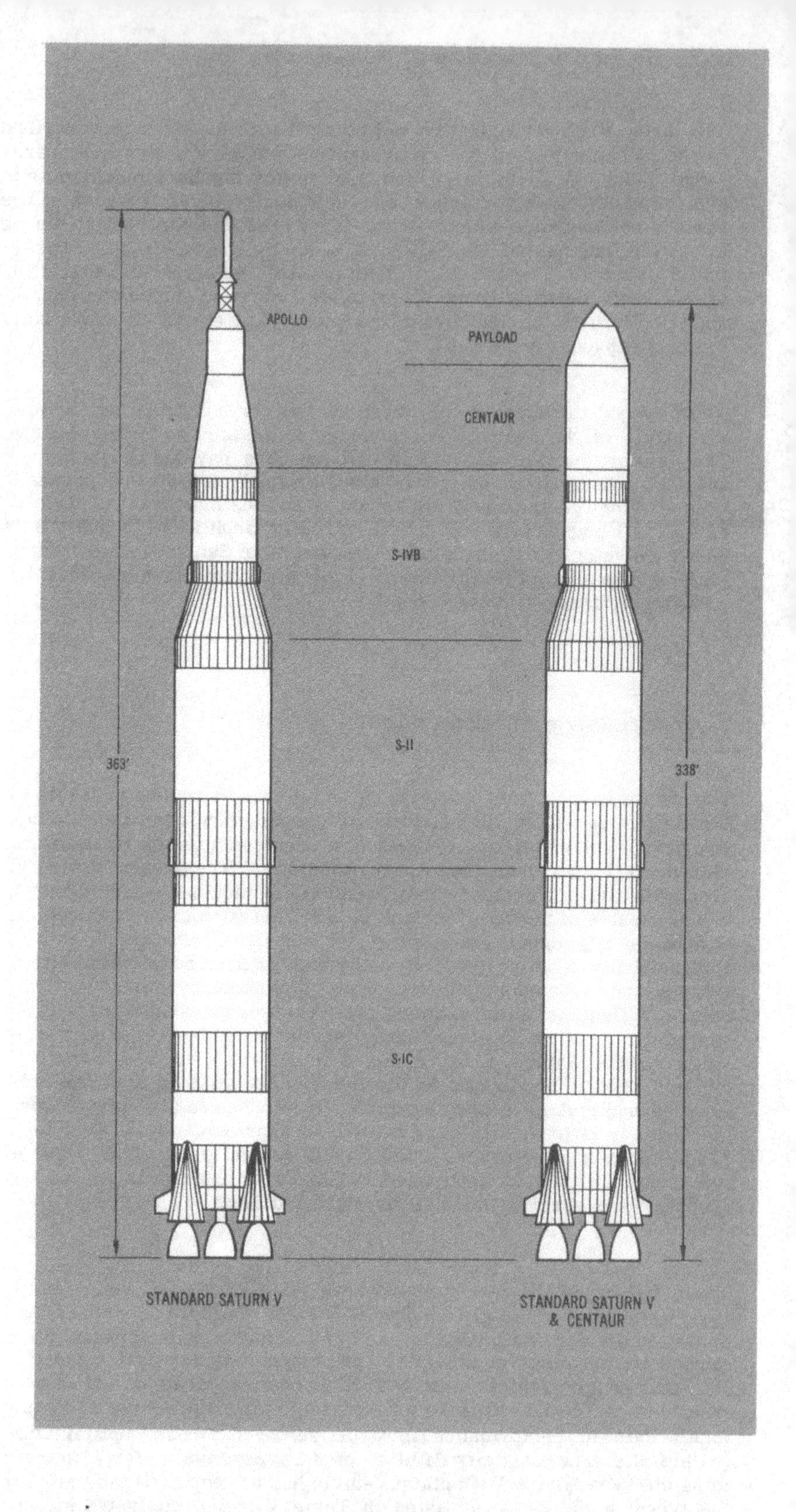

Uprating of the basic Saturn V, as previously described in combination with the Centaur, would result in corresponding payload increases as shown in Table V-II. The basic Saturn V LOR Vehicle thus provides a base for substantial growth potential for accomplishing the high energy missions.

TABLE V-I

VEHICLE WEIGHT SUMMARIES

	Weights Lb.
S-IC at Separation	381,645
S-IC Stage/Residuals	(369,700)
S-IC/S-II Aft Interstage(1)	(1,330)
S-IC/S-II Separation/Start Losses	(10,615)
S-IC Propellant	4,555,003
S-IC/S-II Forward Interstage	9,770
S-II at Separation	100,664
S-II Stage/Residuals	(92,139)
S-II/S-IVB Interstage	(7,468)
S-II/S-IVB Separation/Start Losses	(1,057)
S-II Propellant	969,078
S-IVB at Separation	28,549
S-IVB Dry Stage(2)	(25,708)
S-IVB Residuals	(2,841)
S-IVB Total Usable Propellant Capacity	230,000
Flight Performance Reserve(3)	-
S-IVB Weight Loss in Parking Orbit(4)	3,495
Instrument Unit (8)	3,100
Centaur Shroud	5,933
Centaur Separation Weight(5)	4,564
Centaur Insulation Panels(6)	1,258
S-IVB/Centaur Interstage	630
S-IVB/Centaur Separation/Start Losses	146
Centaur Usable Propellant	29,900
Centaur Weight Loss in Parking Orbit(4)	475
	6,324,210(7) (8)

Note: (1) Two plane separation - 1330 pounds separates with S-IC 9770 pounds is carried with S-II for 30 seconds before jettison.

(2) Includes 204 pounds of jettisoned weight (ullage cases etc.).

(3) Normally computed on a basis using a percentage of vehicle characteristic velocity.

(4) Typical for 4-1/2 hours coast.

(5) Centaur weight includes its own guidance for use during Centaur flight.

(6) Jettisoned after Centaur ignition.

(7) Total weights do not include payload, payload adapter, shroud, or launch escape system.

(8) Based on projected data for SA-513 and subsequent vehicles.

TABLE V-2
TYPICAL SATURN V/CENTAUR HIGH ENERGY MISSIONS EXAMPLES

Mission Category	Target	Possible Mission (1)				(2) Std Sat V/ Centaur Playload	Uprated Sat V/ Centaur Playload
		ITD	ITT	α Deg	V_∞ (3)	Pound	Pound
Planetary Probe	Mars Flyby	1.5	150	2.0	13,379	86,000	97,000
	Jupiter Flyby	5.2	750	-	29,785	36,000	44,000
Comet Intercept	Encke	0.4	100	12.0	30,273	34,000	42,500
	Schwassman-Wachmann	5.5	500	9.5	41,016	16,500	18,000
Asteroid	Ceres	0.2	80	23.0	52,246	6,000	7,000
	Icarus	2.6	200	10.0	36,621	22,000	27,500
Solar Probe	–	0.2	80	-	43,457	14,000	15,000
	–	0.12	76	-	53,223	5,500	6,000
Out-of-the-Ecliptic	–	1.0	200	25.0	45,410	12,000	12,500
	–	1.0	200	35.0	56,641	1,250	3,500
Out-of-the-Solar System	–	40.0	4000	-	44,433	13,000	14,000
	–	40.0	4000	10.0	49,316	8,500	9,000
Libration Point Exploration	Moon-Earth Points	208,000 n. mi.	100 hrs	-	13,600 (Equiv. V_∞)	85,000	95,500

Notes:

(1) Nomenclature

ITD - Interplanetary Target Distance from the Sun (A.U.)

ITT - Interplanetary Transit Time (Days)

α - Inclination of the Flight Plane to the Ecliptic Plane

V_∞ - Hyperbolic Excess Velocity (ft/sec)

(2) Based on SA513 and Subsequent Vehicles

(3) $V_\infty = \sqrt{C_3}$ = Energy Parameter

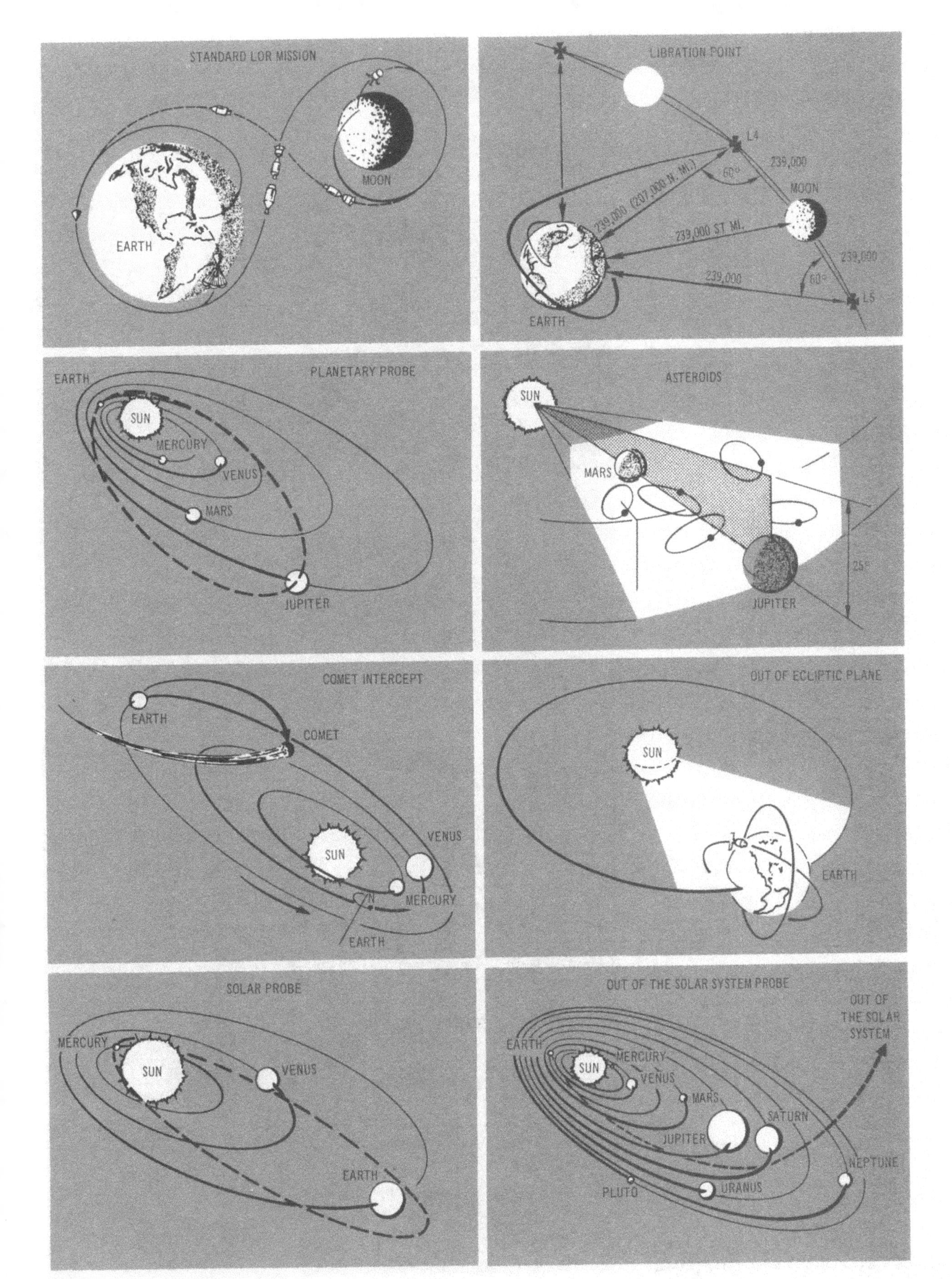
STANDARD LOR MISSION
EARTH
MOON
LIBRATION POINT
L4
239,000
60°
MOON
239,000 (207,000 N. MI.)
239,000 ST MI.
239,000
239,000
60°
L5
EARTH
PLANETARY PROBE
EARTH
SUN
MERCURY
VENUS
MARS
JUPITER
ASTEROIDS
SUN
MARS
JUPITER
25°
COMET INTERCEPT
EARTH
COMET
SUN
VENUS
MERCURY
EARTH
OUT OF ECLIPTIC PLANE
SUN
EARTH
SOLAR PROBE
MERCURY
SUN
VENUS
EARTH
OUT OF THE SOLAR SYSTEM PROBE
OUT OF THE SOLAR SYSTEM
EARTH
SUN
MERCURY
VENUS
MARS
JUPITER
SATURN
NEPTUNE
PLUTO
URANUS

Figure V-3
PAYLOAD VS HYPERBOLIC EXCESS VELOCITY SATURN V/CENTAUR

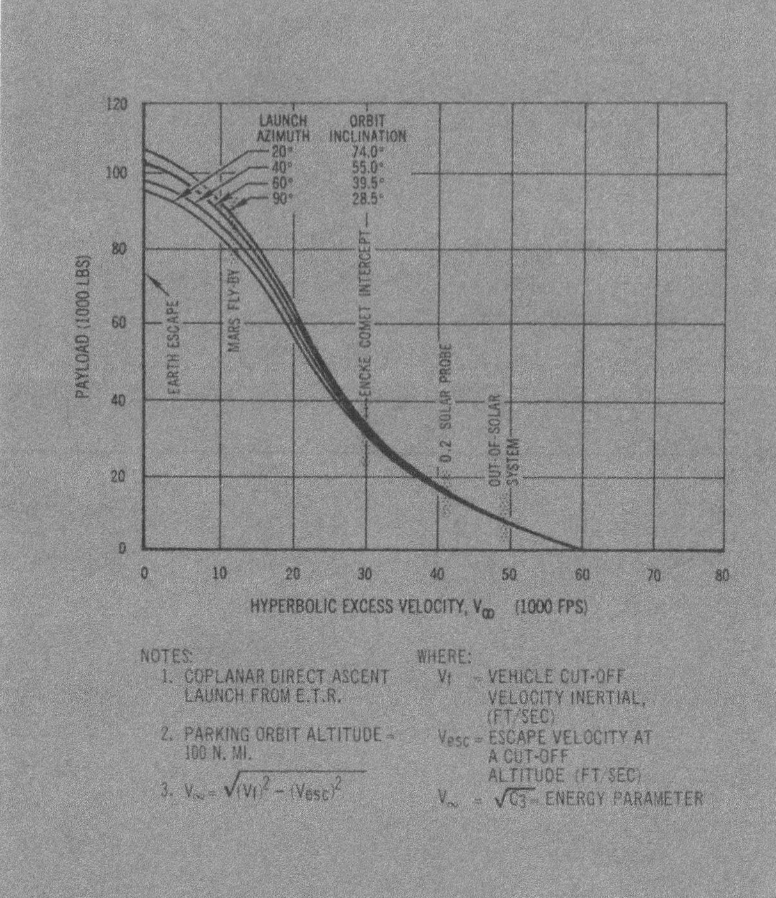

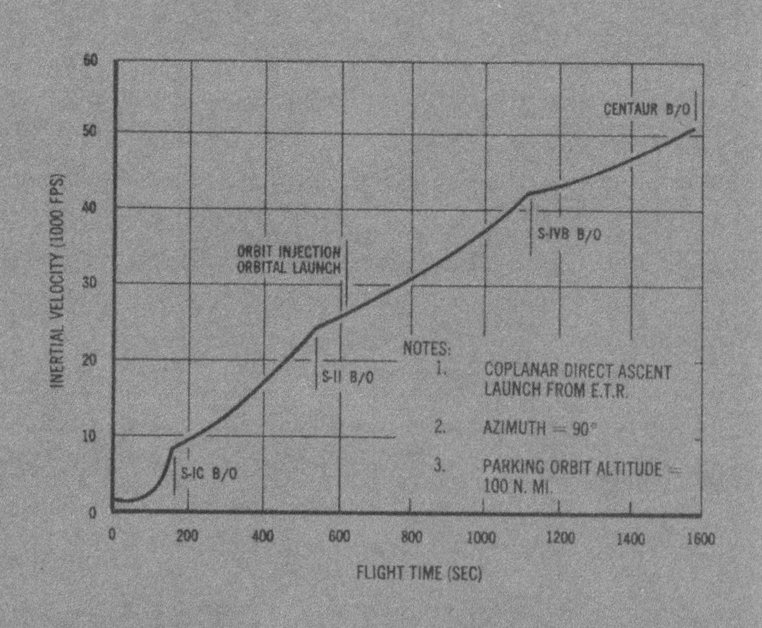

Figure V-4
SATURN V/CENTAUR INERTIAL VELOCITY VS FLIGHT TIME

Figure V-5
SATURN V/CENTAUR
ALTITUDE VS RANGE

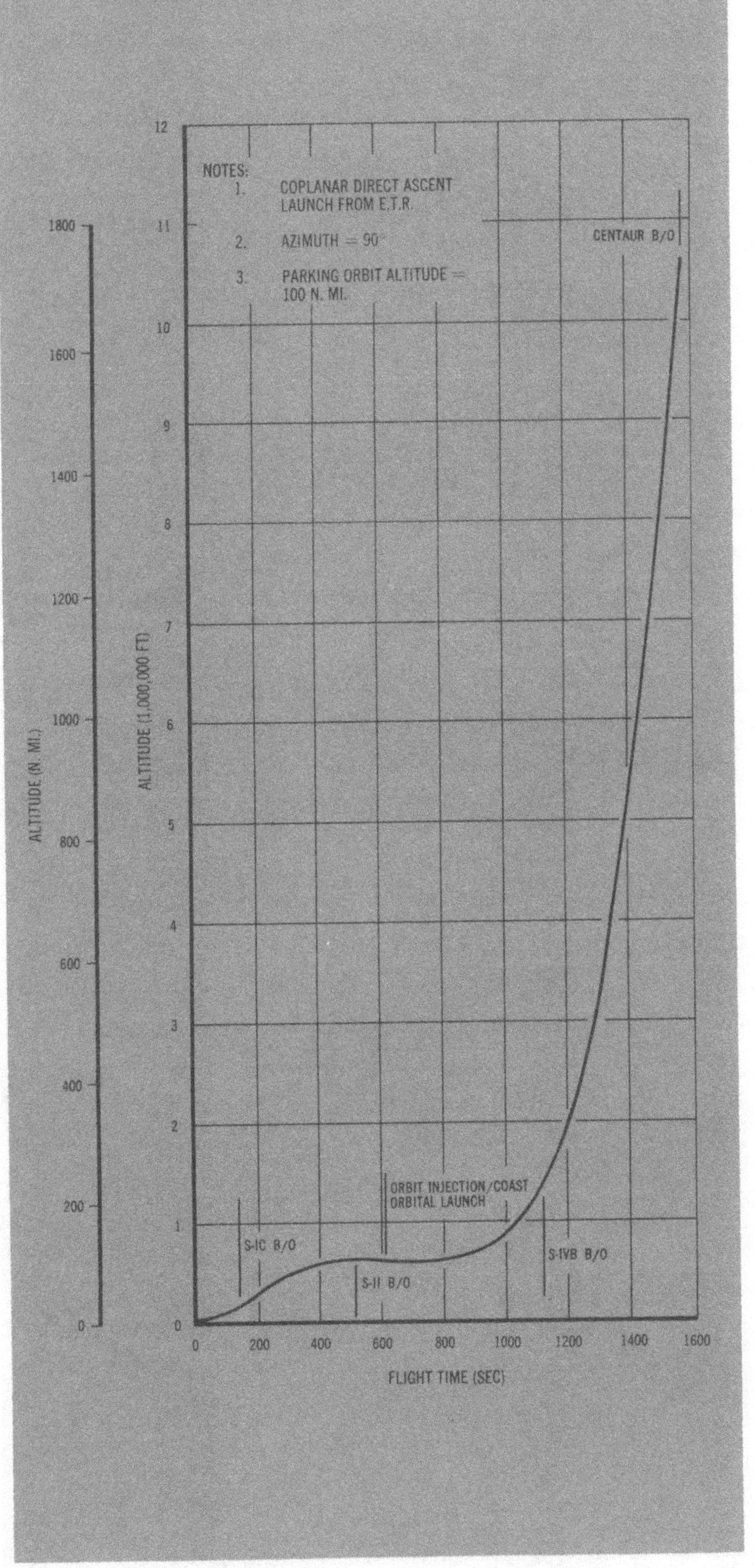

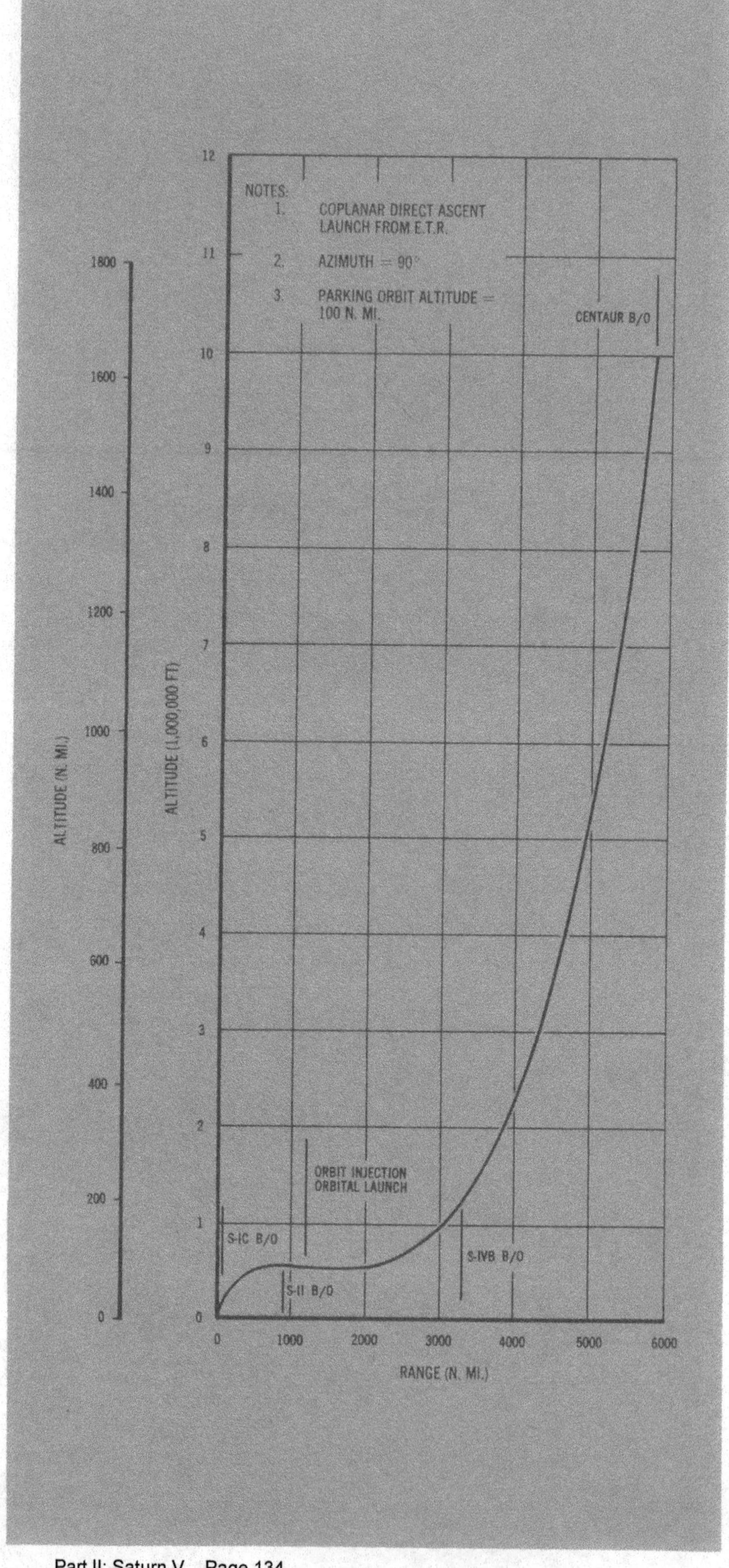

Figure V-6
SATURN V/CENTAUR
ALTITUDE VS FLIGHT TIME

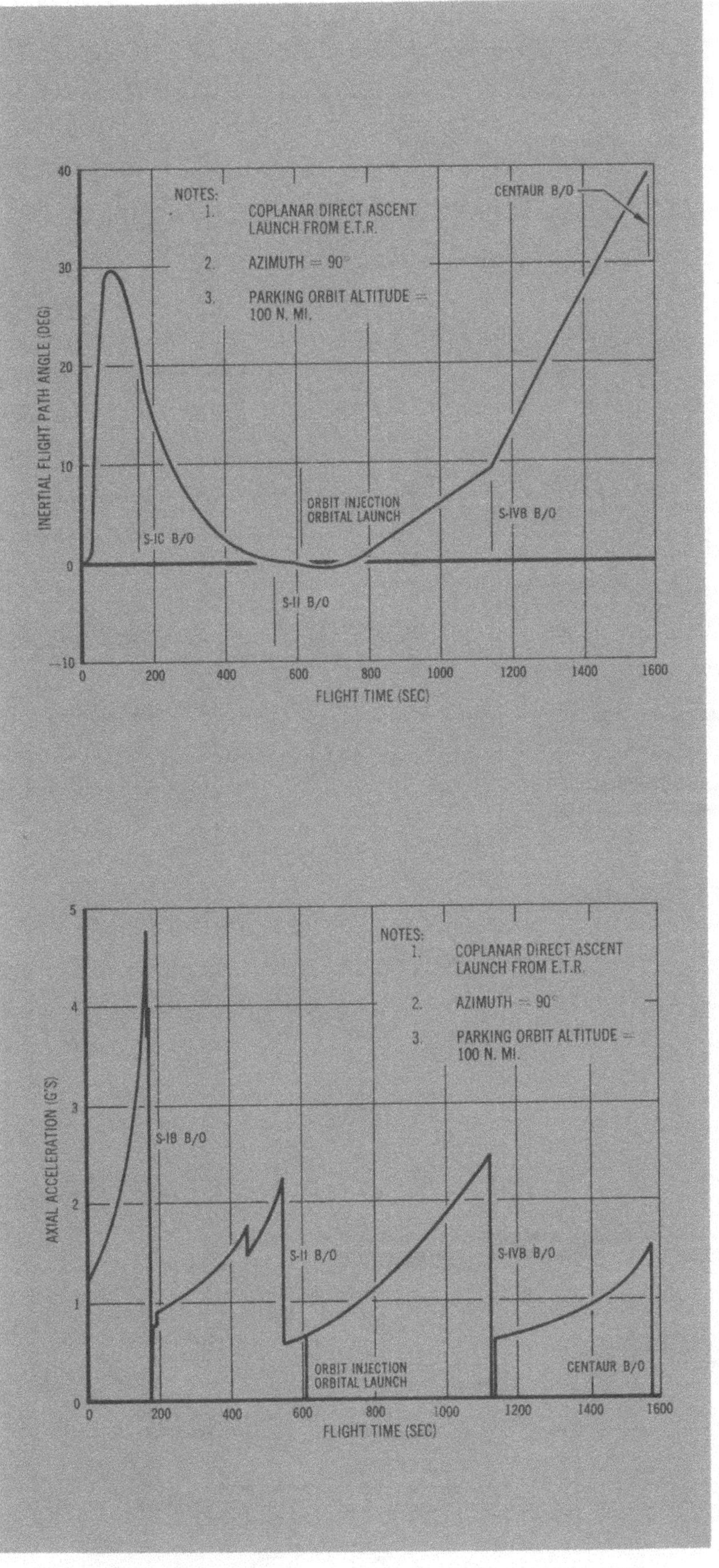

Figure V-7
SATURN V/CENTAUR
INERTIAL FLIGHT PATH
ANGLE VS FLIGHT TIME

Figure V-8
SATURN V/CENTAUR
ACCELERATION
VS FLIGHT TIME

Saturn V can carry:

Experiments in the space sciences . . . . .

Engineering tests that require actual environment of space . . . . .

Prime missions that require a launch system to provide great weight-lifting ability or high velocity.

FOR YOUR CONVENIENCE IN OBTAINING FURTHER INFORMATION RELATIVE TO THIS GUIDE
YOU MAY ALSO CONTACT THE FOLLOWING DOUGLAS – MSSD FIELD OFFICES:

DOUGLAS AIRCRAFT COMPANY, INCORPORATED

SUITE 506
1100 17TH ST. N. W.
WASHINGTON, D. C. 20036
202/628–2556

364 N. ARROWHEAD AVE.
SAN BERNARDINO, CALIF. 92405
714/885–7120

SUITE 13
16501 BROOKPARK ROAD
CLEVELAND, OHIO 44135
216/267–1257

3322 S. MEMORIAL PARKWAY
HUNTSVILLE, ALABAMA 35807
205/881–0611

SUITE 620
333 W. 1ST ST.
DAYTON, OHIO 45402
513/BA 2–6367

69 HICKORY DRIVE
WALTHAM, MASSACHUSETTS 02154
617/899–4115

820 KIEWIT PLAZA
OMAHA, NEBRASKA 68131
402/345–3992

619 N. CASCADE AVE.
COLORADO SPRINGS, COLO. 80903
303/634–5513

SUITE 416
650 N. SEPULVEDA BLVD.
EL SEGUNDO, CALIF. 90245
213/322–5871

SUITE 300
1120 W. MERCURY BLVD.
HAMPTON, VIRGINIA 23366
703/838–2551

16811 EL CAMINO REAL
HOUSTON, TEXAS 77508
713/HU 8–3410

P.O. BOX 400
COCOA BEACH, FLORIDA
305/UL 3–2317

MISSILE & SPACE SYSTEMS DIVISION
DOUGLAS AIRCRAFT COMPANY, INC.
SANTA MONICA/CALIFORNIA

DOUGLAS

GC-1060 PRINTED IN U.S.A.

PROJECT MERCURY

FAMILIARIZATION MANUAL

Manned Satellite Capsule

Periscope Film LLC

LMA 790-1

PROJECT 

APOLLO

lem

LUNAR EXCURSION MODULE

NOW AVAILABLE!

FIRST MANNED LUNAR LANDING

FAMILIARIZATION MANUAL

GRUMMAN AIRCRAFT ENGINEERING CORPORATION • BETHPAGE, L. I., N. Y.

ISBN #978-1-937684-77-8
www.PeriscopeFilm.com

CPSIA information can be obtained
at www.ICGtesting.com
Printed in the USA
BVHW091234270620
582323BV00003B/246